Strategic Highways of Africa

Strategic Highways of Africa

Guy Arnold and Ruth Weiss

ST. MARTIN'S PRESS
NEW YORK

Printed in Great Britain
Library of Congress Catalog Card Number: 76-53953
ISBN: 312-76431-6
First published in the United States of America in 1977.

CONTENTS

MAPS

1 FOREWORD

Developments in various parts of Africa during 1976 continued to high-
light the strategic nature of the continent's highways. Political and
economic strategies changed and then changed again as old highways
were closed or new highways opened. In southern Africa, Mozambique
closed its borders with Rhodesia, forcing the rebel colony to become
still more dependent upon South Africa. In Namibia, following the
debacle of South African intervention in Angola, the new all-weather
military road from Grootfontein to the Caprivi Strip assumed still
greater significance as the Republic geared itself to fight its own
guerrilla war with S W A P O. In East Africa, the Chinese formally
handed over the gleaming, new T A N Z A M railway to Tanzania and
Zambia, while further north an undeclared blockade – denied by Kenya
– threatened to bring about the downfall of the Amin government in
Uganda.

By March 1976 it had already become clear that the détente exercise
launched in 1974 had collapsed, and once Mozambique had closed its
borders with Rhodesia the latter was left with only two rail outlets to
the sea: across Beit Bridge and south to Durban, or through Botswana
and south to the Cape. This fact greatly enhanced the strategic importance of the 400 mile railway from Rhodesia through Botswana. In
April 1976 President Khama said his country was prepared to apply
U.N. sanctions to Rhodesia and close the railway but only if the international community could provide the necessary manpower and
finance to run the railway, for Botswana was in no position to do so
herself. The railway is Botswana's vital carrier: 500,000 tons of imports
and 300,000 tons of exports pass along it and out through South Africa.
There are no railway maintenance facilities in Botswana so that rolling-
stock repairs have to be carried out in South Africa and with that
government's agreement. In terms of the coming war in Rhodesia the
closing of this railway could exert a great deal of extra pressure upon
Salisbury, for in 1975 Rhodesian trade over the line came to a total of
1,380,000 tons: should all this be diverted to the Rutenga-Beit Bridge
line it would add considerable extra strains to the only other outlet
Rhodesia now has.

In July 1976, in his third overseas visit of the year, President Khama
of Botswana went to Peking where, among other things, he was to

discuss the possibility of another Chinese railway in Africa, this time to link Botswana and Zambia. Should this materialise it would assist still further in the process of detaching Botswana from economic dependence upon its neighbour, South Africa.

The riots of June 1976 in Soweto, near Johannesburg, with their near 200 deaths must have brought home to the government in Pretoria as nothing else could the explosive nature of the powder keg upon which it sits. The riots may also have turned the attention of military strategists to the problems of urban control, for these must play an increasing role in the Republic during the 1980s and the single railway line that links Soweto to Johannesburg will be of crucial importance.

The Chinese formally handed over the TANZAM railway to the Tanzanians and Zambians during July 1976: it looks magnificent; the trains run on time; and already it is having an impact upon the lives of the people through whose areas it passes. Its northwards pull will exert economic and political pressures upon events in southern Africa. The railway also created some awkward problems for Tanzania and Zambia: different levels of wages are paid in the two countries (those of Tanzania being the lower) and workers on the regular Zambian network do not want to be transferred to work on the TANZAM if it means a drop in pay. No agreement had been reached at mid-year about TANZAM trains continuing from Kapiri Mposhi to Lusaka and there was a mile-long gap between the terminus of the TANZAM and the old mainline station at Kapiri Mposhi.

The bitter quarrel between Uganda and Kenya that had smouldered on for most of 1976 and then erupted violently following the Israeli rescue of the hijack victims from Entebbe was 'controlled' in a sense, because Kenya sits athwart Uganda's vital road and rail outlets to Mombasa. By the end of July there were repeated rumours of pressures within Uganda to get rid of Amin since, for all practical purposes, it appeared that the economy of the country was grinding to a halt. Fuel stocks were right down and Amin said that Uganda might be forced to "fight for survival". He accused Kenya of mounting a blockade — which Kenya denied — but added that Uganda had enough fuel to enable it to fight for three months. Whatever the truth in the accusations and counter-accusations between Uganda and Kenya the tiny landlocked state of Rwanda whose oil and other essential supplies travel 1,000 miles up to it through Kenya and Uganda had been reduced to a state of crisis and sent ministers to Kampala and Nairobi to ask the two countries to make sure its supplies got through. This east African quarrel also

10

affected supplies for the eastern parts of Zaire.

A more encouraging development was an ECA meeting in Addis Ababa in June 1976 when representatives of six countries agreed to construct a trans-East African highway to link Cairo in Egypt with Gaborone in Botswana. The highway is to run from Egypt through Sudan, Ethiopia, Kenya, Tanzania and Zambia to Botswana and will have feeder roads off into another ten countries.

Highways have a vitality of their own and some of the international ones that serve the more troubled parts of present day Africa have political and strategic roles that give them an importance out of all proportion to their simple economic use as trade routes. Developments in southern Africa, especially during the next few years, are liable to be deeply affected by the pattern of highways now existing or being created. And in no other continent are highways of all kinds — roads, railways, rivers — playing so vital a role in strategy, politics and development as they are in the Africa of the 1970s.

Guy Arnold
Ruth Weiss

August 1976

2 INTRODUCTION

The jet age has tended to discount the concept of strategic highways; yet in Africa they have become more important than ever before. Zambia, Rhodesia, South Africa and Botswana — to name only a few African countries — depend upon highways whose value to them is at least as much political as it is economic. Traditionally, highways are the channels of a country's development, the arteries for its trade and commerce and the movement of its people. They can be much more: means of military confrontation; alternatives to be used as threats in economic conflict with neighbours; or, in the case of rivers, as barriers or boundaries. The strategic nature of highways is especially important in Africa where fourteen of the world's twenty-eight landlocked countries are to be found.

Certain highways — the T A N Z A M * or the Benguela railways — deeply affect the countries through which they pass and profoundly influence the policies those countries adopt. Other projected highways, such as the Trans-African Highway or the Trans-Saharan Highway, have potential political significance for the future.

In the mid-1970s the African continent was on the threshold of great economic and political developments; highways, or their lack, had determined much of its past; and in a communications-conscious age it was very clear that certain highways would play a key role in its future. There are perhaps a dozen strategic highways of the first importance in Africa whose opening up, destruction, or the denial of whose use would fundamentally alter developments on the continent.

Highways have a way of becoming entities with a life of their own: a river can dominate the development of a whole region; a railway can have a personality quite distinct from the countries it serves. The pattern of colonial economic development in Africa from Zambia southwards, for example, has been controlled for decades by the Cape Railway while, historically, Rhodes' dream of a Cape to Cairo railway played a crucial part in the politics of that part of the continent despite the fact that the dream never became a reality. As a result of the development of the Cape railway system, Zambia, Rhodesia and Botswana each looked southwards in economic matters; even major efforts by Zambia to break this pattern in her first post-independence

*The Chinese-built railway from Dar-es-Salaam in Tanzania to Kapiri Mposhi in Zambia, sometimes referred to in the press of those countries as the TAZARA.

12

decade only partly succeeded.

The building of the T A N Z A M railway between Dar-es-Salaam in Tanzania and Kapiri Mposhi in Zambia, and the construction of the B O T Z A M road in Botswana to link her main towns in the centre of the country with Zambia across the Kazungula ferry over the Zambezi, represented deliberate efforts by those two countries – political decisions rather than economic ones – to break a pattern of dependence which had left them largely at the mercy of South Africa. Thus, both these highways – whatever their eventual economic value – were born as the result of strategic-political decisions.

Fourteen countries in Africa are landlocked: it is no accident that among them are the least developed states on the continent. Indeed, in almost exact proportion to the extent of their known or potential mineral wealth these countries are more or less isolated in terms of their communications with the outside world: Niger, Chad, the Central African Republic, Rwanda and Burundi have had few or no resources the outside world has wanted, and strategically they have been unimportant in terms of African and colonial power struggles. As a result they are among the most backward countries on the continent with the least developed communications. On the other hand Zambia and Rhodesia – strategically and economically the two landlocked countries with the most mineral or other resources to develop – also have the best, most complex patterns of highways radiating outwards from them.

Zambia is a classic example of a landlocked country whose politics have been dominated by the question of communications. Historically her development was along the line of rail. After achieving independence in 1964 Zambia was faced with escalating violence in southern Africa as well as the determination of the white regimes, predominantly South Africa, to dominate the region economically. She tried urgently to find support for the construction of the T A N Z A M railway, turning to China only when South Africa's western friends refused to help. The justification for the line was political: to free Zambia of the constraints upon her policies that would inevitably have been placed upon them had she continued to rely solely upon communications through the white-controlled countries. Yet for Zambia the T A N Z A M railway was not enough: by the mid-1970s, after a decade of pressures upon his country's communications, President Kaunda was determined not to rely only upon one route. In consequence, at the start of 1975, even as the T A N Z A M railway was being completed, he announced plans for two new rail lines: one to run from the Zambian Copperbelt across central Angola to join the Benguela railway without passing through Zaire; and one from Lusaka to Malawi to link the Zambian system into the smaller network centred upon Blantyre so as to provide direct rail access to the two Mozambique seaports of Nacala and Beira without the need (in the latter case) of passing through hostile Rhodesia.

The legacy of the colonial communications pattern has a major effect in both negative and positive ways upon modern African policies and much of the development of new highways in the 1970s has been a conscious and deliberate political effort to break these old patterns and create new political and economic alignments.

In his East African days Lugard said, build a railway to where there are people and it will pay: the theory was surely right and in part it motivated the long struggle, at least as far as the Imperial British East Africa Company was concerned, to build the Uganda railway. In the end, however, the railway was built for a variety of imperial and strategic reasons more concerned with Britain's rivalry with the French and her desire to secure the headwaters of the Nile than for the economic betterment of the peoples living around Lake Victoria. As it happens, however, economic developments in East Africa have focused upon it ever since.

In Nigeria, as indeed in all the West African states, the colonial railways — British and French — were built inwards from the coast: either to sources of minerals or to provide the colonial and military authorities with quick access to the peoples of the interior. As a result, all development in the region is vertical: that is, from the coast running inwards rather than horizontal or along the coast so as to line one country with the next. If the latter approach had been adopted a regional communications network would have at least begun to emerge by the time the area achieved its independence. As a result of this pattern West African development has been unnecessarily fragmented so that a series of disconnected enclaves were created that only started to have reasonable communications with each other in the mid-1970s. The consequence of this lack of any regionally planned highways has meant that the states of West Africa have each fostered a degree of insularity that has divided rather than united them: this is especially true between the Anglophone and Francophone states. Such compartmentalized development is one of the more obviously negative results of the colonial past.

Nigerian railways provide a perfect example of how colonial policy worked against the creation of a transport system that would bring about the economic integration of the whole territory. Two lines — one from Lagos and one from Port Harcourt — both go north: their function was, first, political penetration and control; and, second, to bring back to the coast the products of the interior for shipment to the market of the colonial power. Fifteen years after independence Nigeria still did not have a railway line that reached any of her borders to make contact with her neighbours nor was there a line across the southern part of the country from east to west to link some of her main commercial centres.

14

Had Nigeria been planned by the British as a state entity, such lines would have been built; instead, the colony was only developed in terms of Britain's commercial interests so that such a transport network was not deemed necessary. By mid-1975 Nigeria's railways (which had been flimsily constructed anyway) were in a dangerously run-down condition and it had become plain that to play a major part in the new economic community of West Africa — E C O W A S — Nigeria urgently needed a new transport system to link her effectively with her neighbours.

Two of Africa's greatest river highways — the Zambezi and the Congo (Zaire) — have both had negative rather than positive effects upon the countries through which they run, although these negative roles have in their way been of vital importance. Both rivers were major lures to nineteenth-century explorers; in both cases it was hoped they would act as great highways for commerce; and in both cases these expectations were to remain unfulfilled because of the numerous obstacles to navigation (impassable gorges, falls and rapids) on each river. Their subsequent history — after these facts had been established — was widely different yet equally interesting.

The Zambezi, which failed as a highway, became a divide and in this capacity it only really came into its own in the post-independence era: as the demarcation line and geographic boundary between Zambia and white-controlled Rhodesia; as the northernmost military frontier of South Africa where for the latter half of the 1960s and the first half of the 1970s the Republic stationed troops to guard against black nationalist infiltration southwards; and as the symbolic divide between independent black Africa and the white-controlled south. Only after the collapse of the Caetano regime in Portugal during 1974 and the subsequent achievement of independence by Mozambique in June 1975 did the role of the Zambezi in part change; yet also no meeting place could have greater symbolism than the middle of the bridge across the gorge at the Victoria Falls where Kaunda, Smith and Vorster gathered with the Rhodesian nationalists in August 1975 in a vain attempt at détente.

The sheer size of the Congo (Zaire) and the vast stretches which are navigable had the effect of raising expectations out of all proportion to the actual potential of the river as a commercial highway. It was natural to assume that the river would be the highway for goods from and to the interior — the rich Shaba province, for example — where the country's huge mineral resources are to be found. Yet, in fact, from Shaba to the seagoing port of Matadi there are no less than five trans-shipments from river to rail and back again, making the journey slow, tedious and expensive. Only in the 1970s was there serious consideration of building another railway across the centre of Zaire, to break a pattern that had been imposed less by imperialism in this case than by the geographic nature of the river itself.

It is no accident that half the most strategically vital highways in Africa — whether rail or road — are to be found in the southern third of the continent, because it is there that the greatest abundance of mineral wealth was uncovered at the end of the nineteenth and through the first half of the twentieth century. And since the area is also that part of Africa most heavily, longest and most tenaciously settled by an alien white group whose policies have developed into a series of racial justifications for minority political rule, it is also the area most affected by strategic considerations of its communications.

Modern Zambia's communications have been hopelessly intertwined with her policies towards her neighbours and the policies of her neighbours towards her. At the start of 1975 when 50 per cent of her copper — the one commodity vital to her economic survival — was being exported through Angola along the Benguela railway, Zambia might have at last thought that with the imminent opening of the T A N Z A M railway she was finally free of dependence upon routes through the white-dominated south. Meanwhile, in her efforts towards solving the racial tensions of the area without letting them escalate into full-scale warfare, her president had pursued his policy of détente with South Africa. At that stage in his negotiations Zambia's growing, near total, independence of routes through the south strengthened Kaunda's hands in his dealings with Vorster. Eight months later the situation had changed fundamentally: as a result of the developing civil war between rival factions in Angola, the Benguela railway and the port of Lobito were closed to Zambian traffic and so serious had her position become that she was obliged to declare *force majeure* on her copper delivery commitments. There followed a reversal: though the ideological commitment to a peaceful solution in southern Africa may have remained as strong as ever, Zambia suddenly found this almost violently reinforced by the economic necessity once more to look at routes out for her copper, including again those through the south. Here, then, one change of fortune and circumstances in one of Zambia's neighbours, Angola, was enough to upset her policies and force a re-appraisal of her political relationship with the south — for her survival. This change demonstrated the vital role that the Benguela railway has come to play for Zambia and shows how her regional strategy at once has to alter when a single highway is closed to her.

Again and again in modern Africa it is possible to see policies being geared to the choices that existing highways present to a government; and alternatively to see governments consciously embarking upon the creation of new highways in order to alter a direction of political development that an inherited system of communications would other-wise dictate. Such pressures have given rise to many subtleties.

16

Thus, for ten years after independence, Malawi argued that since her strategic highways went southwards and came under the control of the white minority regimes she therefore had no choice but to maintain open and good relations with those regimes – contrary to O.A.U. policy and despite the almost universal hostility of black Africa. She did this with Rhodesia, the Portuguese in Mozambique and with South Africa, being the only African country actually to establish diplomatic relations with Pretoria.

On the other hand Botswana, even though it could argue far more convincingly than Malawi that its geographic position forced it to pursue friendly policies with its white neighbours, maintained only that it could not afford to have bad relations with them although it was obliged to have minimum contact for survival. Beyond that, however, Botswana would not go; instead, she positively looked for ways of breaking a deadlock mainly beyond her control. The most concrete way in which she did this was to embark in 1970 upon the construction of the B O T Z A M road to provide her with a link to the black north and so, to some extent, to lessen her dependence upon routes through the territory of her white neighbours. The psychological value of the B O T Z A M at least matched its economic value.

Thirdly, in southern Africa, the coming to power of Frelimo in Mozambique meant that a radical black government now controls the country through whose territory pass two of Rhodesia's vital rail outlets to the sea to Beira and Lourenco Marques (now called Maputo) and one of South Africa's (to Maputo). In March 1976 the Frelimo government in Maputo closed the country's borders with Rhodesia so denying Salisbury its most used – and cheapest – outlets to the sea; this action by Mozambique fundamentally altered the economic and political balance of the area ensuring thereafter that Rhodesia was completely dependent upon communications through South Africa. The railways in question are of immense strategic and economic importance.

The small town of Jebba in west-central Nigeria had until mid-1975 one bridge across the Niger river: a single-track bridge that carried both the railway and the main road – the one superimposed on the other so that road traffic not only had to wait at either end of the bridge whenever a train was crossing, but also had to wait at one end of the bridge while traffic came from the other side, and vice-versa – from Lagos to the north. A major new bridge for the road was being built across the river in 1975. During the Nigerian civil war at the end of the 1960s it is possible that had the Biafran forces in their most successful push to the west reached Jebba then, effectively, they could have separated the north from the south with possibly incalculable consequences for the war and the future of the country. They failed to do so. All too often, however, communications of vital strategic significance are in fact a matter of one bridge or one geographical point.

Relations between the three East African countries — Kenya, Tanzania and Uganda — have often been difficult and since the volatile General Amin came to power in Uganda in 1971 sometimes dangerously unfriendly. Amin, however, has been far more restrained in his dealings with the Kenyatta government in Kenya than with the Nyerere government in Tanzania. More than once there has been a threatened or actual war between Uganda and Tanzania. Proposed escalations between Uganda and Kenya, however, have always subsided and it has been Amin who has backed down. The explanation for this is simple: Uganda's vital communications with the sea pass through Kenya to Mombasa and not through Tanzania to Dar es Salaam. Whatever the rights and wrongs of any arguments between Uganda and either of her neighbours she can afford — if she must — a collision and diplomatic break with Tanzania, although a comparable one with Kenya would put at risk her communications with the sea along the Uganda railway. It is this fact that has inhibited General Amin in his dealings with President Kenyatta.

A different kind of highway in its effects may be described as a projected developer: that is, a road or railway conceived less in strategic terms (such as T A N Z A M) or to break a past pattern, than in order to bring about closer relations — economic, political and cultural — in a particular area or between particular countries. Two such highways projected and under development are the Trans-Saharan Highway and the Trans-African Highway. In addition, should the E C O W A S states embark upon new highways running east-west along the 'bulge' of Africa, these, too, would fall under the same heading.

E C O W A S — the Economic Community of West African States — was formed at the end of May 1975: it is an ambitious attempt to work towards greater economic and political unity in western Africa, but if it is ever to develop into more than a pious political hope the region urgently needs to create better east-west communications to link its different — mainly coastal — cities and capitals. Much has been said of the strategic importance of the T A N Z A M railway in East Africa; but consider a railway linking Lagos in Nigeria with Dakar in Senegal, cutting across the bulge of West Africa and so passing through some of the most populous and richest territory on the whole continent. It would serve alternating English- and French-speaking areas, and therefore could be one of the most profitable and exciting of all highway ventures.

The politics that inevitably grow up round new highways are themselves an indication of the strategic, political and economic potential of such developments and help shape them. Thus, during 1975 there was discussion in Nigeria as to whether the Trans-African Highway from Lagos to Mombasa or the Trans-Saharan Highway following the ancient caravan route across the desert to Algiers and the Mediterranean had greater economic value for Nigeria. At the time of the debate both roads were more ideas than actuality and both must end (or start) in Nigeria if they are to be meaningful. The eventual outcome of such a Nigerian argument, therefore, and the consequent priority that Nigeria subsequently accords to the one road or the other will clearly have a profound effect upon all the countries through which each of the roads is scheduled to pass.

Strategic highways fall into three general categories. First, there are existing highways that dictate a pattern of relations in the area through which they pass. These include such highways as the Cape railway, the railway from Salisbury to Beira, the Benguela railway and the Uganda railway. Second, there are highways built as a result of political decisions whose main reasoning was to break an existing pattern of communications and so alter the economic and political relations of an area: under this heading come the T A N Z A M railway and the B O T Z A M road. Third, there are new highways which partly reflect the second reason (when they come into existence they may well contribute towards breaking old inherited colonial patterns) but are intended primarily to foster the development of new relations — economic, political and cultural — where none or only limited ones existed before: under this heading come the Trans-Saharan Highway and the Trans-African Highway.

An examination of some of these highways in Africa demonstrates how often such highways are the result of a mixture of political and economic considerations; certainly in the 1970s many of the major political questions and confrontations on the continent were inextricably bound up with such highways.

PART ONE

THE SOUTHERN AFRICAN COMPLEX

The rapidly changing politics of confrontation and détente that characterized southern Africa in the mid-1970s in turn mirrored a pattern of communications that had been established eighty years before. The Rhodesian whites under Smith never intended to hand over political power to their black majority – Smith's notorious remark 'not in my lifetime' accurately reflected the feelings of his white minority. Yet when in 1974 détente became fashionable it was not as a result of any change of mind in Salisbury but because of a change of control over Rhodesia's lines of communications through Mozambique. This in its turn was followed by South African pressures once Rhodesia had become totally dependent upon rail and road communications through the Republic. Rhodesia could not put these communications, upon which her survival depended, at risk by antagonizing Pretoria; so, however reluctantly, Smith began to negotiate – or pretend to negotiate – with his black nationalists.

When Mozambique became independent in June 1975 Rhodesia's two railway links to the sea (from Salisbury to Beira and Bulawayo to Lourenco Marques) were threatened by closure from the new militant left-wing Frelimo government under Samora Machel. Mozambique closed its borders with Rhodesia in March 1976. This meant that from then onwards Salisbury had to plan as though being entirely dependent upon communications through South Africa (including, still, the Cape railway out through Botswana). For the first time the Rhodesian minority government had no communications options to fall back upon when dealing with the government in Pretoria.

In a determined effort to buy time for her own white minority against the increasing pressures from black Africa, South Africa meanwhile changed direction, and almost certainly made a number of secret decisions in 1974–75, following the collapse of Portuguese power in Angola and Mozambique. The essence of these decisions was: to force Rhodesia to come to terms with its nationalists and so rid South Africa of the burden of supporting a colony in revolt on its doorstep; to be prepared, if necessary, to grant some form of independence to Namibia so as to bring about a relaxation of U.N. and world pressures upon an issue where South Africa's position is an essentially illegal one; and to launch another all-out effort at détente and her 'good neighbour' policy towards black Africa in an effort to convince the continent of Pretoria's good intentions. These moves were designed to buy time for Pretoria while the white minority made its internal position more secure, ready to face the siege that it believes must eventually come. The South African Government had clearly decided that the best chance for the survival of white supremacy was to drop its external commitments to Rhodesia and, if necessary, sacrifice Namibia as well in order to reduce the area to be defended to the historic boundaries of the Republic.

The explosive politics of southern Africa are almost all bound up with communications. Not only does Mozambique control two of Rhodesia's sea exits, but for years Beira had served Zambia as her main port of exit for copper, so that the closing of the Rhodesia-Zambia border in January 1973 deprived that country of one of its basic import-export routes. The need to reopen this, especially in the light of the fighting between the liberation movements in Angola in mid-1975, which brought about the closure of the port of Lobito and the Benguela railway to Zambian copper, made Kaunda's efforts to bring about détente all the more vital to Zambia. Before independence Machel and the leaders of Frelimo had been outspoken in their condemnations of South African and Rhodesian policies, stating that they would close the borders and have no relations with the white regimes once independence had been achieved. After June 1975, however, the new Mozambique Government was notable for a time for its silence on the issue of the borders and especially on the question of South Africa's continuing use of Lourenco Marques (Maputo). Mozambique might, for ideological reasons, ban South Africa from using its railway and port, but it is one of the poorest countries on the continent and the South African traffic contributes a substantial proportion of annual revenue. The immediate post-independence excuse for no action was that Frelimo had first to ensure her control of the country.

In October 1975 Smith made one of his major slips when on a British television programme he blamed current setbacks to détente upon South African interference. There was uproar (some officially inspired) in the South African press, not least because South Africa was looking for an excuse to force a reluctant Smith into a corner and make him come to terms with his nationalists. Smith was then obliged to fly to Pretoria and make a humiliating public apology to Vorster for his remarks (though he later denied having done so). He did so because by then he and his regime had been reduced to absolute dependence upon communications through South Africa and should the latter refuse to allow Rhodesia to use its railways to the Cape and Durban then indeed U.D.I. would be likely to be over in a matter of weeks rather than months. The row was a classic illustration of how control of highways gives the whip-hand to one country over the policies pursued by another.

Politics and communications in southern Africa cannot be separated. The stridency of Lesotho's denunciations of apartheid, for example, under the government of Lebua Jonathan from 1970 onwards was in direct proportion to the real feeling of helplessness of a small and economically weak country entirely surrounded by South Africa and therefore entirely dependent upon the latter's communications system. Smith's U.D.I. itself was taken in 1965 in the knowledge that his communications through South Africa and Mozambique were then safe while his readiness to negotiate with the African National Congress — or factions of it — ten years later grew with the uncertainties surrounding these routes.

South African troop dispositions on the Zambezi from the later 1960s through to their withdrawal in mid-1975 during the détente exercise, were the forward defences of a system all of whose communications radiated outwards from the Republic itself; a major communications gap — to take troops by road to the exposed Caprivi Strip — was rectified in the early 1970s by the building of two roads whose purpose was almost entirely strategic (see pp. 41 and 162).

The politics of the south revolve round the Witwatersrand, the heart of Afrikanerdom and the powerhouse of the South African economy; the southern African communications network was built up largely to service the Rand or to by-pass it and the history of the growth of this communications system illustrates clearly how strategic highways have come to play so large a part in the politics of the area.

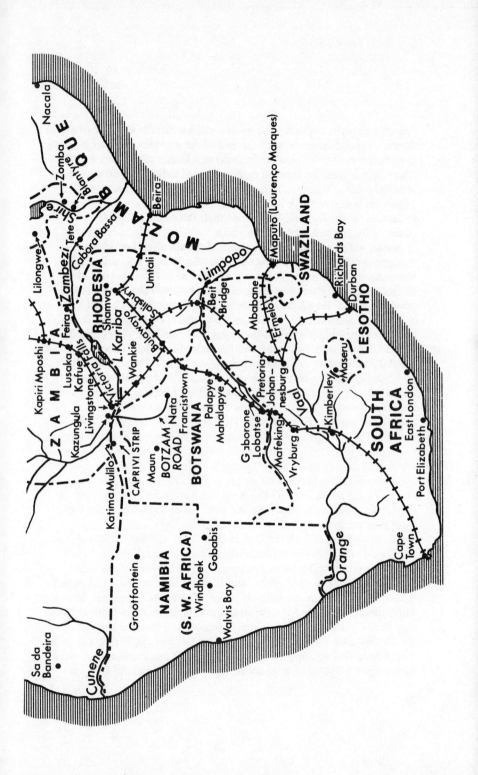

The crucial period in which the basic communications network of South Africa was established came at the end of the nineteenth century during the unprecedented mineral boom in diamonds and gold which turned South Africa from an imperial trouble-spot of indifferent value into an imperial prize – still equally troublesome – of enormous worth. The 1886 gold boom of the Witwatersrand led the colonies to borrow money in London as they frantically pushed their railways northwards to compete for the rich trade of the Rand.

South Africa entered the rail age in 1860 when the first steam train chugged slowly between Durban and the Point along 1.87 miles for five minutes on June 26th of that year. By 1875, 154 miles of rail existed in the Cape and five miles in Natal; by 1890 the Cape line had grown to a total length of 2,403 miles which by then had passed beyond the diamond city of Kimberley to reach Vryburg. Five years later the remaining portion of the Natal line to the Witwatersrand was opened, a total of 3,545 miles. By that time the golden Witwatersrand, the sixty-mile stretch of the original section of the Golden Reef of which Johannesburg became the centre, was linked by rail to the ports of Delagoa Bay in Mozambique, Durban, East London, Port Elizabeth and Cape Town. It was the location of the ever-growing known deposits of natural resources which dictated the spread of the lines – that and the configuration of the country. The absence of navigable rivers made the introduction of railways essential. The size of the investment, the absence of domestic savings and the rapidity of the economic growth of this area made state ownership inevitable. After the discovery of diamonds in 1867 the miners at Kimberley needed material food, tools and timber. Ox waggons were not enough, hence the swift push into the interior of the railway from the Cape.

Between 1885 and 1895 the railway mileage of South Africa almost doubled. During 1886 the Transvaal made an approach to the Cape for a customs union and a railway extension from Kimberley but was rebuffed. When the value of the gold finds on the Rand was confirmed it became the turn of the Cape to be rebuffed by the Transvaal. The Portuguese then started the construction of a railway line from Delagoa Bay to the Transvaal border which it reached in 1889 and thereafter the Netherlands Railway Company started to build the Pretoria stretch of the line. This was opened in 1894. Kruger in the Transvaal appeared intent upon building up a separatist republican axis behind Delagoa

Bay and between the Zambezi and Vaal rivers. He refused to join a railways and tariff union with the Cape and Natal on imperial terms; in their turn the railways from the south appeared to be threatened by the Delagoa Bay line, the more so as Kruger obstructed the lines from the south crossing republican territory to Johannesburg. By 1895 the Rand was providing Kruger with the economic power that would enable him to exercise a controlling influence in South African affairs.

Thus gold and the developments that followed it led the Transvaal to redouble its efforts to establish its own communications with the sea while the English colonists did their best to prevent this: 'This deadlock over a railway and customs union brought on a scramble for territory from the Transvaal, northward as far as Lake Tanganyika, and eastward to Swaziland, Kosi Bay and Delagoa Bay.'[1] The Cape colonists concentrated upon their expansion northwards through Bechuanaland to the Zambezi and beyond. As it was Bechuanaland already hemmed in the Transvaal on the west and Rhodesia was to do so in the north, although first (for the British) there was a Portuguese threat in the area to be dealt with.

Natal was far weaker than the Cape, yet it too was pushing its communications north to cut Transvaal's access to the sea and was busy petitioning London for help: in 1886 the Natalians and their governor, Sir Henry Bulwer, urged the Colonial Office to check the republicans and reserve the colony's hinterland. Natal wanted Zululand for its own natives. Forty-two MPs signed a memorial to prevent the Zulu falling under the Boers; Britain then prevented the Transvaalers in Zululand from reaching the sea. In 1888 Tongaland was declared a British sphere of influence; to conciliate the Boers, however, a quarter of Zululand was yielded to the new Republic.

In all these manoeuvres and counter-manoeuvres the Delagoa Bay line — Lourenco Marques to the Rand — was the key railway. Kruger was determined to release the Transvaal from colonial commercial monopoly; to do this he had to have the railway to Delagoa Bay through Portuguese and not British territory. In 1887 McMurdo's Anglo-American syndicate began work on the Portuguese end of the line; after abortive conferences between the two sides in 1886 and and 1887 it became clear that Kruger meant to stop the colonial railways entering the Transvaal and, if possible, the Orange Free State as well, until the Delagoa Bay line was finished. Rhodes said of this railway in 1886: 'If the Delagoa Bay Railway is carried out, the real union of South Africa will be indefinitely deferred.'[2]

For their part the Cape Dutch farmers urgently needed a railway connexion with the Rand gold fields and after 1886 Hofmeyr joined Rhodes in preaching a railway policy that would join Table Bay with

Johannesburg and Delagoa Bay. It was Kruger's determination to have nothing to do with such schemes that had brought about the Rhodes-Hofmeyr alliance. The next move was a decision of the Cape and Natal leaders to buy control of the Delagoa Bay railway so as to keep their share of the Rand prosperity. From 1887 onwards they began to bid for McMurdo's company. When asked for British support Knutsford at the Colonial Office said he would 'be glad to see the railway in colonial hands'[3] but the Prime Minister would not interfere. In any case the British government could not give any guarantees for the Delagoa Bay line since it passed through Portuguese territory.

At this stage Lord Salisbury thought the Transvaal and Portugal in conjunction could only seriously subvert British supremacy in the area if they received major German aid; this he did not think they would get. The attempt to buy the railway failed for lack of imperial support; then in June 1889 the Portuguese government in agreement with the Transvaal confiscated the line and carried it forward to the border. Rhodes tried to buy the Transvaal section but Kruger bolstered the company with state funds. Rhodes, who habitually thought that money could do anything, tried to buy the southern province of Mozambique. Efforts to buy the line were to continue into the 1890s.

If Kruger was winning the railway battle to the south, the imperial forces were taking over control to the north. Salisbury sent Harry Johnston to sort out the problems of the Shire valley and the Portuguese advance up the Zambezi, and Johnston urged the takeover of Zambesia – now Zambia – justifying such a takeover because it 'may one day serve as a link between Egypt and the Cape',[4] for by then he had accepted Rhodes' Cape to Cairo dream. But in May 1889 the imperial outlook changed. Rhodes was then in London to form his British South Africa (B.S.A.) Company. The terms for the new company were sent by Knutsford to Salisbury on May 1st. Rhodes was prepared to amalgamate and administer south and north Zambesia and Nyasa as well – foreshadowing by more than sixty years the Central African Federation.

Meanwhile the Cape line – refused access to the Transvaal – reached Vryburg from Kimberley in 1890 and then the Bechuanaland Railway Company took it to Mafeking. In 1892 Kruger was in a position to borrow in London the money needed for the Delagoa Bay line which was the Transvaal's shortest distance to the sea. At that point Rhodes feared to lose control: he saw the possibility of the Transvaal from the south invading his new colony of Rhodesia and then Zambesia beyond it. Proposals were put forward to offer Swaziland to Kruger to keep him quiet, and in December 1894 Swaziland passed under republican control. Ripon, the Colonial Secretary, then made sure that this concession did not give the Transvaalers independent

26

access to Kosi Bay and to do so he annexed Trans-Pongoland in April 1895 and later (May) re-annexed Amatongaland. *The Economist* at the time said: 'We should be the last to prevent the access of the Boers to the sea on the grounds of trade. We merely wish to prevent the Boers from acquiring a political right to a sea frontier . . . which . . . may be embarrassing to us in our dealings with Foreign Powers.'[5] If the Boers could not be excluded on one pretext, another was as good.

Repeatedly during the 1890s Rhodes tried to purchase Delagoa Bay from the Portuguese so as to cut off Kruger's line to the sea. He failed because there was no diplomatic support for him from London, since first Lord Salisbury and then Rosebery were reserving all their political efforts to keep Germany quiet over Egypt, Sudan and East Africa. An 1890 ultimatum to Portugal led to the British moving into the Shire valley. By June 1894 Rosebery told Rhodes that he must give up all hope of acquiring Lourenco Marques or the railway since the Germans would not allow it. That was the year the railway was completed, and Natal then joined Transvaal in a commercial alliance against the Cape and the Orange Free State customs union; as a result Kruger allowed the Natal railway through to the Rand, promising it a third share of the gross traffic to Durban, another third being reserved for Delagoa Bay. By the end of that year, when the Sivewright agreement* of 1891 expired, Kruger regained control of freight rates on the Transvaal section of the Cape line and diverted Cape traffice – then 80 per cent – to the other two lines so that the Cape became a suitor for one third of the Rand's traffic. As German support for the Transvaal mounted it became Britain's turn to warn Germany that for Delagoa Bay to change hands could mean war, while a Colonial Office memorandum of March 26th, 1896, stated it was vital that the Delagoa Bay Railway did not fall into the hands of any power other than Portugal or Britain. There was a last chance for Britain to get control of the Delagoa Bay Railway in 1898 when Portugal was bankrupt and Germany and Britain were both looking at ways of bailing her out with loans. But this came to nothing.

Rhodes' dream of a Cape to Cairo railway had implications far beyond South Africa. Born in 1853, a millionaire by the time he was twenty-five and Prime Minister of the Cape by 1890, Rhodes saw the Cape to Kimberley line as only the start of an imperial line that would run right up a British-controlled Africa. The line reached the areas of wealth in the Orange Free State without impediment and was pushed promptly to Mafeking, but thereafter Rhodes had to fight for each additional mile. He pushed the railway through the Kalahari desert and

---

*The agreement allowed the Cape line to be extended to Johannesburg and in exchange the Cape government had control of the freight rates on the northern section of the line for three years.

scrub of what is now Botswana to the Zambezi, but progress became increasingly slow and the line was destined never to reach Cairo. Meanwhile Rhodes sent his pioneer column equipped with guns and protected by the police of his new B.S.A. Company to found Rhodesia and after the defeat of Lobengula and the vacating of Bulawayo to the settlers Rhodes saw the way clear — as he thought — for his railway to continue northwards. When he died in 1902 it had reached Bulawayo, 1,625 miles north of its starting point at Cape Town.

The railway was continued north across the Victoria Falls into Zambezia (Zambia). It had then been Rhodes' intention to move the railway from the Zambezi across German East Africa and Uganda to connect with the Sudan-Egyptian line, but Germany refused permission for Rhodes to build across their East African colony, having no wish for a British-controlled line through their territory. Instead the line was to be diverted to the Copperbelt of Northern Rhodesia and eventually it crossed the border of the Belgian Congo to link up with the Benguela railway running westwards to the Atlantic. This part of the line was to play a crucial part in the development of modern Zambia, providing the main means of moving her copper out — until the mid-1970s — and also being the focus for developments along the line of rail.

If the railway mileage of South Africa almost doubled between 1885 and 1895, it doubled again during the period 1910 to 1970. It had reached 7,063 miles in 1910 when it was brought under a unified administration with headquarters in Johannesburg. In 1973 railway lines in South Africa and Namibia totalled 13,905 miles:

Cape	5,514.63
Transvaal	3,694.49
Orange Free State	1,675.36
Natal	1,525.19
S.W.A.	1,462.62
Private lines	32.86
Total miles	13,905.15

Namibia — South West Africa — had been linked with the South African system at Upington in 1915. In 1922 it was added to the integrated administration of South African Railways and Harbours.

Today South Africa's transportation organization comprises a rail network, commercial harbours, a domestic and international air service, extensive road transport services and two pipelines for petroleum and related products. The 1910 arrangement which came into effect on May 31st that year combined into one enterprise the previously autonomous organizations of the various provinces.

Working agreements exist between South Africa and Mozambique and Rhodesia. Overborder working of trains is facilitated in accordance with the agreement, rolling stock is shared, light repairs are made in the country which holds the coaches at the time these are needed, and revenue from rail traffic between one country to the other is allocated and accounted for.

The ports are an important factor in South Africa's international relations. It is through the railways and ports that the South African Government can control Smith's Rhodesia: if South Africa chooses, she can throttle Rhodesia's trade, especially since March 1976 after sanctions had been applied by Mozambique and the traditional Rhodesian outlets to Beira and Maputo had been blocked. Since there is a uniform system of control of railways and harbours, the move could easily be made by the South African authorities.

Durban is the largest port in Africa and handles 58 per cent of South Africa's seagoing traffic, about thirty million tons of cargo a year. It has repair facilities, off-shore terminal and single-buoy mooring for oil tankers, a sugar terminal (privately owned), pre-cooling stores, extensive shed facilities and other amenities for commercial shipping. Port Elizabeth was developed for the handling of mineral ores: it has an annual productivity of about 8 million tons. The port of East London is South Africa's only river port. It handles citrus fruit and is ideal for the accommodation of smaller craft. Walvis Bay, which is part of the Cape Province though geographically in Namibia, is a fishing centre and important to the Republic as its main point of entry on the west coast. Richards Bay has been developed to take the load off Durban. It is about 121 miles north of Durban and the deep-water harbour that has been constructed there will facilitate a further expansion of South Africa's international trade. It is designed to handle bulk items such as coal. The first berths opened in 1976. During 1972–3 South African ports handled 52 million tons of traffic compared with 21.6 million tons ten years before.

South Africa is the world's third largest fresh-fruit exporter and here it is vulnerable to world opinion: it is easier to carry through a fruit boycott than one against strategic materials such as mineral ores which the western countries need. If action against South Africa because of its race policies is ever to take the form of international boycotts, the ports will play an important role in trying to keep South African commerce going. The country has its own fleet, Safmarine, with refrigerated vessels which would have to run the gauntlet of any boycott. To be effective a boycott against South Africa in reality would mean a blockade in the Indian and Atlantic oceans of the country's major outlets – and there is no possibility of such international action in the foreseeable future.

South Africa is the dominant power in the region: she represents formidable political and economic force to all her neighbours. In terms of her race policies she is unacceptable as an ally to anyone in Africa apart from white Rhodesia. For reasons of geography and power some countries — Botswana, Lesotho and Swaziland — are forced to trade with her and use the communications she controls; one country — Namibia — is directly (if illegally) controlled by her; Malawi alone of black African countries chose to open diplomatic relations with her even though communications between the two are indirect and must first pass through Mozambique and Rhodesia. Other African countries — whatever O.A.U. resolutions they may have subscribed to — carry on overt or covert trade with her: Zambia as a result of the past still has substantial trade with the Republic especially in terms of equipment for her mining industry; even distant Zaire trades with South Africa, and the railway system that originates in the Republic goes as far as Zaire through Rhodesia and Zambia; while Mozambique, independent in June 1975, holds the key to fresh changes in the area's communications pattern. As it is Mozambique is so poor and reaps such considerable income from the transit of South African goods through the port of Lourenco Marques that, despite ideological differences which could hardly be greater, she may still find she has no option but to continue to trade with South Africa and to permit the transit through her territory of the Republic's imports and exports.

South Africa's economic objectives are easily understood. Her coastline has few natural harbours although there are rich fishing waters off the coast, while poor arable land in a large part of the country made it essential that the thrust into the continent continued. With no navigable rivers the transport system inevitably became dependent upon rail and road, though air transport has come to play a part. Over a third of the rail mileage of sub-Saharan Africa exists in the the Republic.

The South African Railways and Harbours Corporation is the largest commercial undertaking in the Republic. It employs over 250,000 people. During the depression in the 1920s and 1930s it became a vehicle of social security, employing thousands of homeless, ill-educated poor whites — mainly of Boer descent — and this too helped the development of the network as it exists today with many sidings and tracks that were built at that time now in full use, though then they were hardly economic. The staff is preponderantly Afrikaans.

The South African road network is also an extensive one, covering over 220,000 miles, 10 per cent of these being tarred. Private long-distance haulage is controlled by the Motor Carriers Transportation Act which was established to protect the monopoly of the State railways.

Although the emphasis for the railway system in the 1970s is upon the maintenance of existing lines, new links are still being built such as that between Vryheid (Sikame) and Empangeni (127 miles), Groveput and Vogelstruisbult (30 miles), Broodsnyersplaas and Ermelo (58 miles), Empangeni and Richards Bay (11 miles) and Arnot power station and Wonderfontein (14 miles).[6] In view of the low cost of coal the main sources of power are steam and steam-generated electricity. The ultimate aim is to have electric traction on all important lines and diesel traction on the lines which are long distance from the coalfields.

Natal was the first province to have electrified lines, the first one having come into operation in 1926; today 30 per cent of the entire network is under electric traction. Mainline passenger and goods services are operated by electric locomotives. The modernized signalling system is electric and employs electronic techniques. Computers are used to assist with the control of the large stations. C.T.C. − centralized traffic control − systems have long been in use and new ones are being installed with the aim of extending it to all main lines. The distance under C.T.C. was 791 miles in 1975 and was to be increased by 1976 by an additional 808 miles.

The South African Railways and Harbours Corporation is constantly expanding and wants ever more rolling stock for its growing traffic. The rapid population growth, the stagnation of the country's agricultural production and the lack of large internal markets are all partly responsible for the Republic's efforts to thrust deeply into the black hinterland of the continent: to do so the railway has become a necessary tool; it served the imperial ambitions of Rhodes in this way and is a weapon in the armoury of Vorster as he pursues his policy of détente. Indeed, South Africa's railways have always been a major factor in its economic development. The country's infrastructure has been strengthened by the extensive railway network which in turn has helped expand foreign trade as huge tonnages of mineral resources are easily transported from inland points to the main export ports. The main lines are being strengthened and modernized to increase railway productivity; rolling stock is constantly being purchased − most of it is made in South Africa − to replace obsolete equipment, and newly developed specialized wagons include ore carriers, refrigerator cars, wagons for oil and explosives and others for coal, grain and timber. The passenger section of the railways − at least that which caters to the whites − has been updated and the number of coaches in service in 1972−73 was 8,570 as opposed to 6,695 ten years earlier.

South Africa's white population has a high ratio of motor cars − 318 to every 1,000 people, which compares favourably with figures of 310 per thousand in Canada, 269 in Australia, 255 in Sweden − and 424 in the United States. This high density of cars is the reason for the low growth of white passenger traffic on the railways: urban passenger rail traffic for that group showed only a 1.4 per cent increase a year from 1962 to 1972, while non-white traffic increased over the same period at an annual rate of 7.4 per cent. On the other hand goods traffic has increased steadily: in 1972–73, 119.6 million tons were transported compared with 84.8 million tons in 1962–63. Ores and minerals – the main items being manganese, chrome and iron ore – totalled 8 million tons exported during 1972–73.

This formidable railway network means that South Africa has an infrastructure whose value to her poorer neighbours is obvious, and they find it difficult to resist using it even when they can find alter-natives. Strategically the railway is South Africa's most important asset, able to move troops quickly in sufficient numbers to reinforce those travelling by air or to transport equipment to threatened borders. It is also an instrument that can be used to supervise the movement of the black population: in this connexion the urban railways are likely to assume a greater strategic significance in the coming years as black-white tensions increase.

One of the most important effects of this South African railway network, with first-class ports at the railheads, is that the whole direction of trade for the area, including Zambia, Botswana, Rhodesia and Zaire is pulled southwards. And one of the most important political and physical developments on the African continent in the 1970s – the building of the T A N Z A M railway – was another direct result of the South African system in the negative sense that it in order to break free of its pull Zambia and Tanzania had to have an alternative. The T A N Z A M will help change both the political and economic alignments of the area. Despite this particular – though spectacular – railway development the Rand remains the commercial centre of power in southern Africa; control of the railways is still in the hands of the Republic as far as Rhodesia is concerned; and at the end of 1975 and into 1976 South Africa appeared to be confidently pursuing its policy of a détente whose possibilities were at least in part based upon the Republic's control of such a railway infrastructure. A major change in the power structure in the region could be brought about should the Frelimo government of Mozambique decide to end the Republic's use of the railway to Maputo since that line remains as strategically and commercially valuable as ever.

Notes

1. Robinson and Gallagher, *Africa and the Victorians*, (Macmillan, London, (1961), see page 214.
2. Rhodes speech, Cape House, 1886.
3. Quoted in Robinson and Gallagher, *op. cit.*

4. *ibid.*
5. *The Economist*, May 11th, 1895.
6. *South Africa 1974*, Official Yearbook.

4   THE THREE ENCLAVES: BOTSWANA AND THE
   B O T Z A M ROAD

As a result of the First World War South Africa in 1919 became the
mandatory power under the League of Nations for German South-
West Africa and this, including the Caprivi Strip, meant that she all
but encircled Bechuanaland. Swaziland — annexed and used as a pawn
in the power manoeuvres of the 1880s and 1890s — emerged finally as
a British colony, a small enclave three parts surrounded by South
Africa, sharing a fourth border with Mozambique. Basutoland, in
similar style, emerged as a British colony ̀entirely surrounded by South
Africa. For most of the period 1910 to 1961 (when South Africa left
the Commonwealth) Pretoria repeatedly exerted pressure upon
Whitehall in the hope that Britain would hand over control of these
three High Commission territories to her so that they could be
incorporated into the Union. They were not, but emerged instead to
a precarious independence in the late 1960s: Botswana and Lesotho
in 1966, Swaziland in 1968. Much of their subsequent history has
been concerned with their struggle to maintain a minimum of
independence from their all-powerful neighbour. To do this requires
some form of control — for them — over their communications.
   Botswana was in the strongest position. At least the Cape railway
passed through her territory into Rhodesia and in theory she could
send goods that way and so through to Beira and Mozambique. In
commercial terms this made no sense and in any case after U.D.I. in
Rhodesia in 1965 Botswana found herself virtually hemmed in by
white supremacist states with the sole exception of her disputed
river crossing border with Zambia. In her determination to break as
far free as possible of South African control she embarked in 1970
upon the B O T Z A M road project. When Rhodesia gains independence
under black majority rule, Botswana will be in an infinitely stronger
position; and in the perhaps longer-term event of independence for
Namibia where, for example, there is the possibility of another railway
from Botswana across to Walvis Bay, then the combination of that, the
Rhodesia outlet and the river crossing into Zambia would mean that in
theory, if not in effective economic practice, Botswana could be
independent of transit through the Republic.
   Lesotho's case differs greatly. She has no alternative to transport
through South Africa which surrounds her entirely. There are roads
only in the lower regions of her largely mountainous country: 236
miles of main roads and 575 miles of secondary roads. A railway links

Maseru, the capital, into the South African system across the border but otherwise there is no railway within the country itself although the line from the capital is to be extended northwards to Leribe, the industrial area. Lesotho's only direct link to the rest of the world that will not be controlled by South Africa will be by air – when Maseru becomes an international airport, a move that was under consideration with U.N. advice in 1975, but even then the degree to which she will continue as a geographic and political captive of South Africa will remain almost total.

The case of Swaziland is somewhat better, for her border with Mozambique is only a short distance from the sea. Her railway runs from Ngwenya across the country and through Mozambique to Maputo which is her principal outlet. It is a short railway yet of exceptional geopolitical importance in the convoluted situation of southern Africa, for under an independent Mozambique Swaziland could have her major export-import route passing through a friendly black state and need no longer feel she is dependent upon either the political or economic goodwill of South Africa. This railway, therefore, has great significance in giving Swaziland a psychological independence from her powerful white neighbour. South Africa rejected the idea of extending her railway into Swaziland during the 1960s as being uneconomic; however as the political climate altered in the mid-1970s she changed her mind and would like to see a link that renders Swaziland less tied to a radical Mozambique.

The existing line from Ngwenya in western Swaziland to Goba in Mozambique and thence to Maputo is 137 miles long, cost R79 million and was opened in 1964. It is, therefore, a recent addition to the high-ways of the area. It carries 95 per cent of Swaziland's exports – some three million tons of freight annually – so that the little kingdom does not rely upon the Republic. Swaziland's asbestos ore, however, is carried into South Africa by overhead cable for rail trans-shipment.

The changes taking place in southern Africa constantly alter the political and economic significance of existing highways. Political independence in Mozambique and then in Rhodesia and Namibia when it comes will help the other peripheral countries – Botswana, Zambia, Malawi and Swaziland – to break away further from Pretoria's control.

Botswana and the BOTZAM road

The Protectorate of Bechuanaland could be said to have been pro-claimed to secure for Britain, at the behest of Cecil Rhodes, a strategic highway that would by-pass the Boer Republic of the Transvaal and give access to Matabeleland and points further north. It was a part of Rhodes' grand strategy for an imperial highway from the Cape to

Cairo. Developments in Bechuanaland were to follow the eastern strip of the country which was the only part that interested the British as a road to the interior: it was the route to be followed by Rhodes' pioneers to Rhodesia. As with so much imperial history, the positive aspects of a decision were matched by the negative ones. Essentially Bechuanaland was taken over as a Protectorate in 1885 for reasons of communications: positively, to provide the route to by-pass republican Transvaal; negatively, to prevent control of this northward passage falling into the hands of the Germans who were then pressing eastwards from South-West Africa.

A great deal of manoeuvring and intrigue surrounded the building of the railway through Bechuanaland. The British High Commissioner at the Cape, Hercules Robinson, wanted to extend the railway from the Cape through the new Protectorate to secure Matabeleland without having to depend upon the Boers, then firmly entrenched and antagonistic in their two republics of the Orange Free State and the Transvaal. At the same time the Cape leaders and merchants wanted a railway with the shortest route to Johannesburg, so they could tap the riches of the Rand — and that meant a line passing through the Boer republics. The President of the Transvaal, Kruger, played into Robinson's hands, however, when he refused to allow a railway from the Cape to the Rand. This allowed Robinson, working with the Sprigg Ministry at the Cape, to split the opposition at the Cape and carry a Bechuanaland Extension Bill to allow the railway to by-pass the Transvaal. At that point Kruger offered to admit a colonial line to the Rand if the Cape government dropped the Bechuanaland extension.[1]

After the signing of the Rudd Concession by Lobengula in 1888 (whereby the latter granted a monopoly of minerals in his kingdom) Rhodes urgently needed the railway to further his plans for northern expansion. He was greatly helped in 1889 by Kruger's obstinacy in refusing to allow a line through to Johannesburg which enraged the Cape Boers, Kruger's normal allies, so that they were prepared to support Rhodes in his plans for a Bechuanaland railway. Rhodes had a major ally in Robinson who bombarded the Colonial Secretary with arguments for the establishment of a Chartered Company which, he said, would save the British taxpayer money and prevent the area falling to the Boers. In May 1889 Knutsford (the British Colonial Secretary) recommended Rhodes' proposals for the British South Africa Company to Lord Salisbury; subsequently the government gave the company its charter, first and foremost to ensure that the colonial grip on southern Africa could be tightened. Rhodes himself said that the government made him promise to build the Bechuanaland railway before giving him the charter: 'It was upon the strength of this pledge that my application was favourably regarded by Her Majesty's

Government, and that the British South Africa Company has been granted a Charter.'[2]

In the fast-moving politics of southern Africa at the time it was remarkable how the reluctant imperialism of a few years previously that had only just acquired Bechuanaland developed into a conviction on the part of Salisbury and the Colonial Office that the line was crucial. They believed it was necessary both to develop Bechuanaland commercially and to strengthen the Cape's hand in dealing with Kruger so as to bring about a railway and customs union (one that modern Botswana is less than happy to belong to) needed to bind together the Cape, Bechuanaland and the supposedly rich north. It was thought, moreover, that the railway would break down the republican axis then being built between Johannesburg and Delagoa Bay. When it was built the railway took the same line from south to north over some 400 miles as the old 'Missionaries' Road': it was the preservation of this road from Boer encroachments that had become a crucial part of the Bechuanaland Protectorate. The railway became a necessity as soon as the settlement of Mashonaland and Matabeleland got underway in earnest following the first settler column north in 1890. In 1892 Alfred Beit formed a subsidiary of the B.S.A. Company to build the railway from Vryburg to Palapye in return for a government subsidy and smaller subsidies from the B.S.A. and Tati companies. The company was to have a large block of land in the new Protectorate.

At the end of 1892 Rhodes went to London to press the railway on the Colonial Office. The result was the 'Railway Despatch'[3] in which the Secretary of State approved the railway and gave the B.S.A. Company several additional privileges. Local officials were ordered to do their best to persuade chiefs and the people to grant concessions to the Company while the High Commissioner was to investigate all other claims to concessions in the Protectorate and no others made after the Charter were to be recognized.

Once started the company worked at speed: the Vryburg-Mafeking section of the line was opened to traffic early in October 1894 while the line to Bulawayo was completed by the end of 1897, the last 400 miles taking only 400 days. The railway has remained the communications backbone of Bechuanaland (Botswana) ever since. It is the main route for imports and exports and carries the overwhelming proportion of internal tonnages. For modern Botswana, however, there is the complication that the railway belongs to and is operated by Rhodesia and until 1974 it was the only Rhodesian railway route out through South Africa. The line was built essentially as a means of communication between the Cape and Rhodesia and only incidentally as a means of developing Bechuanaland. The railway may be Botswana's main communications artery but it is also a limiting one — determining the direction of its trade and forcing it into a mould that can only be effectively broken by the establishment of new routes. It has been claimed that: 'The main arteries of Botswana's communications system are the railway and parallel road which run through the populated eastern part of the country and connect with the South African and Rhodesian systems. Severance of these routes would effectively divide Botswana, or isolate it.'[4]

The railway is 398 miles long. From Mahalapye (its mid-point in the country) it is a distance of 850 miles due west to Walvis Bay, while distances to Durban, Port Elizabeth, Maputo and Beira are all approximately 686 miles. Virtually all Botswana's imports and exports have to use the longer route to the Cape − at commensurately higher cost. In terms of possible new developments a Botswana rail link to the sea through Namibia could become a reality when that country achieves independence from South Africa. The Namibian rail system has been extended to Gobabis only 62 miles from Botswana's western border and this connects directly with the port of Walvis Bay.

Botswana is one of Africa's fourteen landlocked countries: the inevitable complications of such a position are for her made far worse because she is surrounded by white-controlled minority regimes to whose policies and race philosophies she is adamantly opposed. As a result of this situation and most especially in reaction against the dominant and pervasive power of her neighbour, South Africa, Botswana has pursued a policy − in so far as she is able − of maximizing her contacts with the independent north, mainly that is, with Zambia.

As Seretse Khama has said:

'Our only common frontier with independent majority-ruled Africa is a narrow disputed one with our sister Republic of Zambia.' He also said: 'And these states [the white-ruled ones] are not only our neighbours; for historical and geographical reasons, which are none of our choosing, we also trade with them. Our economies have long been closely interlinked and we depend on their transport systems for our outlets to the world. The main railway connecting South Africa and Rhodesia crosses Botswana and we rely on it not only for our trade with the outside world but for internal communications. It is wholly owned and operated by Rhodesia Railways.'[5]

Thus the communications system upon which Botswana relies is un-favourable to her interests: partly because in fact the railway she has to use is owned and operated by Rhodesia; and partly because − in comparison with Rhodesia − Botswana is only a marginal user of the railway and so goods for Botswana from the Cape, for example, are accorded a low priority.

Against this historical background — itself one of strategic communications considerations — the idea of developing a major road connexion with Zambia across the Kazungula ferry over the Zambezi became of major importance during the troubled years of the early 1970s as black-white tensions throughout the area were steadily escalating. The route from the south to Kazungula is an old one. Before 1904 the ferry was the main entry point into western Zambia (Barotseland) across the Zambezi. King Lewanika declared it to be the only legal point of entry. The ferry service was by canoe and in 1873—74 waggons, for example, were carried across by two canoes. In 1885, however, an iron vessel, the *Holub*, was operated at the ferry by an Englishman, Westbeech.[6] After the railway and bridge were established across the river at Victoria Falls, the ferry at Kazungula survived mainly as a route for migrant labour on its way to and from South Africa from Barotseland and Angola, and as a point for swimming cattle across the river from Ngamiland to the Livingstone market.

But the ferry took on a new lease of life in the 1960s with the development of tourism and the building of the Chobe River hotel in Botswana's Chobe gamepark. Although for many years the road through Bechuanaland to the ferry was little more than a sand track it was nevertheless constantly used both as a point of entry into Barotseland and the Batoka plateau of what is now Zambia and, as Lord Hailey said, it 'has been maintained by the Witwatersrand Labour Association for the transport of labour from Barotseland and Angola to the Rand Mines.'[7] This fact was ignored or overlooked by Vorster during the controversy of 1970 when he opposed the building of the Nata-Kazungula road by the Americans. It is also a matter of record that during the colonial period — prior to Botswana's independence in 1966 — South Africa put considerable pressure upon the colonial authorities either to maintain the road to the ferry or to permit the Republic to undertake the task. This again was conveniently forgotten in the events of 1970.

The 1960s saw a general escalation of black-white tensions throughout southern Africa: the guerrilla wars in Angola and Mozambique; the fortifying of the Caprivi Strip by the South Africans, including the construction of a major military air-base at Katima Mulilo; the subsequent escalation of guerrilla activities in the Strip from 1966 onwards by S W A P O (the South West African People's Organization); and finally the growth of guerrilla activity in Rhodesia which had ensured that by 1970 several thousand South African troops and para-military police were committed to guarding the Zambezi border between Rhodesia and Zambia. These various confrontations meant an increasing flow of guerrillas and refugees in both directions: refugees from South Africa and South-West Africa came through Botswana and then headed north, using what came to be called the 'Freedom Ferry' at Kazungula on their way into Zambia. This refugee movement provided an important reason for South Africa's opposition to any improvement of the route.

Meanwhile, in the late 1960s an American loan to Botswana was used to construct the Francistown-Maun road upon which stands Nata, the turning-off point for the road north to Kazungula. American interest in Botswana at that time (1969) was not great, but two things were to increase it: the fact that the Chinese had embarked upon the building of the T A N Z A M railway from Dar es Salaam southwards to Zambia (the *Uhuru Railway*), and the U.S.A. wished to be involved in its own version of a Freedom Road; and the fact that the Nixon administration wanted to demonstrate to its black pressure groups that it was active — on the right side — in southern Africa. Both reasons acted as spurs to persuade the U.S.A. to become involved in the B O T Z A M road. Thus, when they were seeking a Botswana aid project in 1969 and had been presented with a list of 24 possible projects, (listed in order of priority) the Americans chose number 17, which was the B O T Z A M road, then rated far lower in terms of importance than it was subsequently to become. The road had the right political appeal in terms of the criteria U S A I D was applying to the area. At that time the Nata to Kazungula stretch of road was to be the northern end of a heavy all-duty road 625 miles long from Lobatse in the south through Gaborone the capital to Francistown and then across to Maun with the northern 180 miles branching off from Nata.

The road was in the balance in the early 1970s because of soaring costs and political pressures as well as a degree of footdragging by the Americans whose southern Africa policy under Nixon and then Kissinger increasingly became one of appeasing Pretoria. Work on the design of the road was only completed in May 1972 but until the end of that year, although the strategic importance of the road for Botswana was clearly increasing, its economic possibilities still appeared to be minimal.

It was only in September 1972 that an agreement was signed in Gaborone whereby U S A I D was to provide a $12.6 million loan repayable after a ten-year grace period over thirty years at 3 per cent interest: this was to cover the cost of a gravel surface road from Nata to Kazungula. It was then estimated that it would take 32 months to construct and there were to be two feeder roads: one of 3.7 miles, the other of 40 miles. The smaller of the two spurs went from Panda ma Tenga to the Rhodesian border and into the Wankie Game Reserve.

40

The other spur ran from Kazungula itself along the south bank of the Chobe river to Ngoma where there is a bridge across the river due south of the South African air base at Katima Mulilo in the Caprivi Strip. This section of road was designed to facilitate traffic from Rhodesia to Katima Mulilo and at that time, using the old gamepark road, there were in the region of 2 to 4 trucks and 12 pick-ups a day passing along it for the Caprivi Strip. This supply line for South African Caprivi operations is an embarrassment to Botswana and appears to have been included in the total road plan by the Americans as a sop to Pretoria's sensitivities. By 1973 the cost of the road had already increased from $12.6 million to $16.6 million; the road will be completed in 1976.

On her side Zambia built an eight-mile stretch of road to the ferry linking it with the tarmac road from Livingstone into western Zambia. According to the terms of the U S A I D loan, the construction had to go to an American company and was awarded to Grove International. As part of the general agreement Zambia was to provide an improved ferry service and north bank facilities for dealing with heavy transport. The U S A I D loan was to cover 186 miles of 'graded gravel highway' plus 44 miles of feeder roads. Ironically, by the mid-1970s, it had become plain that while the new heavy-duty tarred road was being completed between Lobatse and Gaborone and then on to Francistown, costing up to R37,500 a mile (due to be completed in 1977), the 188 miles of untarred road from Nata to Kazungula were going to cost about the same as the tarred section.

The B O T Z A M road was the largest single project in the 1973–78 development plan and represented the biggest side effect from the new mining developments.[8] With crucial mining expansion taking place and the continuing uncertainties of southern African politics to contend with in the years to come the road was intended to ensure the supply of bulk fuel needed for the mining operations. By mid-1974 the Zambian side of the project was complete: the road was tarred and two ferries were in operation.

The key to the road has been Botswana's determination to break out of South Africa's economic stranglehold upon her – a result of both history and geography. The Zambian connexion across the ferry represents the only geographical means of such a release – hence the enormous strategic rather than economic importance of the road. The building of the road became a physical necessity as a real link between Botswana and the north. South Africa – from her own point of view – was right to protest at the road, for it was bound to lead to a lessening of her influence and control. In any case there were certainly technical questions about the precise border relationship between the four territories at the crossing – Botswana, the Caprivi Strip, Zambia and Rhodesia – especially as two of the territories were in international dispute at the time anyway: Caprivi as part of South-West Africa was deemed by the U.N. as well as the International Court at the Hague to be illegally held by South Africa while Rhodesia was a British colony in revolt against the Crown. The point at issue was the precise location of the common frontier at the confluence of the Zambezi and the Chobe rivers.

In terms of the whole Botswana road programme, Gaborone had rejected a South African offer to tar the Lobatse-Gaborone section of the road, having no wish to be beholden to Pretoria. Instead, she went ahead with the Nata-Kazungula plan: small as is present trade between Botswana and Zambia, the long-term possibilities are more encouraging. Development of Botswana's Makgadigadi salt pans could provide Zambia with both salt and soda ash; and there is the longer-term possibility of a rail link at the ferry that would bring in the Botswana system to connect it with the TANZAM railway to the north and conversely – if the link across Botswana to Namibia comes about after the latter's independence – the chance for Zambia to develop yet another exit, to Walvis Bay, through Botswana. In the mid-1970s, however, the road represents a vital lifeline for Botswana with independent black Africa, one which can free her from the isolation of being surrounded by white minority regimes.

So much for the project itself and the Zambian connexion that Botswana wished to develop. The road, however, was only to get underway after significant political pressures had been mounted by South Africa to prevent it. During 1970 South Africa was to deny that Botswana had any border at all with Zambia and therefore to argue that she could not build the road to the ferry. Both Zambia and Botswana deliberately adopted a low profile approach to South Africa and did not answer her notes about the road. Once the row had got underway, however, the U.S.A. dropped hints that continuing South African opposition to the building of the road could result in Washington raising awkward questions about the South African presence in the Caprivi Strip: this was doubly illegal not only because of the U.N. decision on South Africa's continued retention of the Mandate but also because even according to the old League Mandate South Africa had no right to station troops in the Caprivi Strip at all. Apart from this British colonial records showed that the ferry had existed and been in regular use prior to colonial times.

The South African involvement in the road was far more complicated than would appear from her 1970 rejection of the concept of a common Botswana-Zambia frontier. Prior to Botswana's independence in 1966 South Africa had tried to persuade Britain to give her permission to improve the road linking Caprivi with Rhodesia, but the colonial authorities had not agreed. The South African Ministry of

Defence had, indeed, surveyed the route itself. The American decision to build the road was itself clearly a political one and in part South Africa was reacting against this: when the project was announced Vorster protested. In retrospect it would seem he was not briefed by either his military or foreign affairs advisers unless his protest was a highly sophisticated bluff (of a kind South Africa has never yet demonstrated), for what emerged from the events of 1970 was an American readiness to build an additional feeder road from Kazungula to Ngoma whose only real beneficiary was to be the South African military establishment in the Caprivi Strip at the Katima Mulilo base.

It was in March 1970 that U S A I D announced it would support the project and build the road to link Livingstone in Zambia and Francistown in Botswana. South African reaction was swift and in April Pretoria sent a note to Gaborone to the effect that Botswana had no common frontier with Zambia: this was during the 1970 South African election when electoral pressures as well as strategic ones were at play. Although at first Botswana did not want to reply to the South African allegations and played down the whole affair, pressures forced the government to make a statement on April 14th, 1970:

> The Government of Botswana has received a note from South Africa which expresses the South African view that Botswana has no common frontier with Zambia.
> The Botswana Government sees no reason to change its existing view that Botswana and Zambia have a common though undefined boundary at Kazungula.
> As far as the proposed Nata/Kazungula road is concerned, the Botswana Government takes the view that this in no way alters the status quo since it simply improves access to the Kazungula ferry, which has been operating unchallenged for many years. Not only will the road open up possibilities of trade with Zambia, but it will make possible the development of a hitherto inaccessible region of Botswana.
> Botswana has no reason to believe that the South African initiative is likely to affect the attitude of the U.S.A. Government.

There were complex legal and survey problems relating to the ferry crossing for the converging boundaries are defined in terms of notional points relating to the Chobe and Zambezi rivers and the rivers have from time to time changed course. Furthermore, there exist no precedents for any easy settlement of any dispute since it is the only meeting place of four national boundaries in the world. No one, including the South Africans once their election was over, appeared to want a major row and presidents Kaunda and Khama, especially, worked together in a quiet joint approach. In June 1970, in answer to a question in Parliament, Vorster said:

> I have no jurisdiction over links that might or might not be built between independent countries. If the question implies whether that road link will pass over South African territory that, of course, is a different matter and the South African government has made its position very clear to the Botswana Government. I have nothing further to add to that.[9]

It was, in fact, a South African climb-down. That September Dr Hilgard Muller, the Foreign Minister, said that South Africa did not wish to interfere with the internal affairs of its neighbouring states and he pointed out that for many years there had been a ferry across the Zambezi at that point. Muller went on to claim that South Africa's reaction to the road and ferry question had related to the way Botswana had defined its borders for the purpose of its U.N. membership by showing a common border with Zambia and he claimed that the South African note was 'purely a juridical problem. It had nothing to do with the proposed road.'[10] At the time Botswana and Zambia were speaking only of a road and motorized ferries and not of the possibility of a bridge. The BOTZAM project did not envisage a bridge over the Zambezi, so the precise location of the frontier — somewhere in the middle of the river — was not in question. Clearly South Africa wished to pre-empt the possibility of a bridge being constructed to replace the ferry.

The row about the road and ferry arose because of South African fears of Botswana turning northwards, achieving better access to Zambia and so becoming less dependent upon the Republic. Apart from this, there was the escalating guerrilla activity that by then South Africa had come to take as a serious threat: Pretoria had no desire to see any improvement of the so-called 'freedom route'. When the row became public it was clear that South Africa had no case — as she knew full well — and to persist would have been to draw unwelcome attention to her illegal presence in Namibia and especially to that of her troops in the Caprivi Strip. Moreover, with the inclusion in the project of the 40-mile Ngoma stretch of the road that would facilitate supplies passing from Rhodesia to Katima Mulilo, South Africa was in part reconciled. Even with this addition South Africa saw the road as a security risk and the ferry as a means of escape for refugees, while in reverse it could become a conveyor belt for guerrillas moving southwards. What is certain is that the road will act

44

as one more factor making it easier for Botswana to withdraw from South African influences and pressures.

The American position as the main donor providing the loan for the project has remained ambiguous. On the one hand South Africa did not want to oppose the U.S.A. building the road since there is heavy American investment in South Africa. On the other hand the U.S.A. maintained it was helping Botswanan development as the road would open up the country and provide transport links for the development of gypsum and soda ash production from the Makgadigadi pans.

The road is both strategic and economic: on completion it will have a number of effects upon the area as a whole. In the first place it will raise the condition of the people in northern Botswana and south-west Zambia — two regions that so far have been little favoured by development. To some extent at least it will bring them out of isolation and link them more to the main stream of events in their two countries. The road will stimulate activities along the stretch from Kazungula to both Francistown and Maun, and at Maun Botswana is to build its second abattoir (expected to go into operation in the early 1980s) to serve the cattle industry of the northern part of the country while Maun itself will develop into a modern township. Nata is to have a telecommunications centre and will become the halfway house between the ferry and Francistown which is the start of the route south. The road will encourage trade between Botswana and Zambia and (whatever eventual settlement is achieved in Rhodesia) reduce Zambia's long-term dependence upon Rhodesia Railways. Immediately, Botswana's concern with the road is as a cattle route to Zambia and thence to Zaire. Other effects — again in the future — will be substantial: to link Botswana more closely in with Zambia and Tanzania so that in effect she need be less closely integrated with the South African customs union and economy. There is the possibility, already mentioned, of an eventual rail link using the crossing to tie in the Zambian system with any Botswana line across Namibia — but such a development could only come after Namibian independence when a bridge across the river at Kazungula could no longer be disputed by South Africa.

There will be a number of benefits for Zambia: the road was in any case given a major economic and strategic boost in January 1973 when Smith closed the Rhodesia border with Zambia so giving the route immediate and urgent importance. Since then Zambia has done some re-routing of essential mining equipment from South Africa as well as chilled beef across the ferry while in the future she could import salt and soda ash that way for her domestic and mining consumption — that is, when Botswana can break the present South

African monopoly and start to operate the Makgadigadi pans. One of the aspects of the route that is strategically most important for Botswana will be its use for the bulk transport of oil and other products from Livingstone down to the Shashe project to feed the new mining complex there. It could also be an import route for maize and sorghum from Zambia, for Botswana is only self-sufficient in these foodstuffs one year in ten. More generally, it is a link that will foster growing trade, tourism and other relations between Botswana and Zambia and give to Botswana the feeling of being linked to independent black Africa despite being surrounded by her white-controlled neighbours. Even when both Rhodesia and Namibia achieve majority rule – perhaps by 1980 – the link northwards will still be vital to the development of the area.

Although it got off to a slow start at the beginning of the 1970s and at first came fairly low on Botswana's list of priorities, the road became increasingly important during 1973 with the escalating guerrilla wars and the growing political confrontation of the area. Its importance was again enhanced when South Africa refused to accept the idea of an international treaty guaranteeing Botswana, Lesotho and Swaziland freedom of transit for their imports and exports. Further, the possibility of an Arab oil embargo against South Africa made it imperative that Botswana could obtain her oil requirements by another route than that through the Republic, and the B O T Z A M provides the answer to this problem, since oil for Botswana can be pumped down the pipeline to Ndola from Dar es Salaam and then be taken by road to Kazungula. Suddenly, then, in 1973 the B O T Z A M was no longer viewed as economically marginal; instead, it had become a strategic highway of considerable importance and priority.

Notes

1. For a detailed account of the beginning of the Bechuanaland Railway see, Robinson & Gallagher, *Africa and the Victorians*, pp. 231–42.
2. *ibid.*
3. See Anthony Sillery, *Botswana* (Methuen, London, 1974).
4. See Z. Cervenka (ed.), *Landlocked Countries of Africa* (The Scandinavian Institute of African Studies, 1973).
5. 'Botswana and Southern Africa', an address by President Seretse Khama to the Foreign Policy Association of Denmark, November 13th, 1970.
6. See A.N. Parsons, *Economics of the Zambia – Botswana Highway*, *Enterprise*, No. 3, 1974.
7. See Lord Hailey, *An African Survey* (O.U.P., London, 1956).
8. See A.N. Parsons, *op cit*.
9. See *Rhodesia Herald*, June 8th, 1970.
10. See *Africa Contemporary Record 1970–1971*, ed. C. Legum, p. B473.

5 THE ZAMBEZI

During the 1960s the Zambezi came to be regarded as the dividing line
between independent black Africa to the north and white-controlled
Africa to the south. It was heavily patrolled, and witnessed an increasing
number of military incidents. Rhodesian, South African and
Portuguese troops held it on one side; members of the various liberation
movements crossed it from north to south when they were ready to
take guerrilla action into the white-controlled territories. When the
détente exercise got underway at the end of 1974 and through 1975 an
uneasy peace settled along the river; it was a peace that both sides knew
could be broken at any time.

The Zambezi river is one of the great waterways of Africa. Regarded
in the nineteenth century by European explorers and imperial
adventurers as a major gateway to the interior and as a means of
penetrating to then unknown parts of the continent, it was never to
live up to these expectations since, like most of Africa's rivers, it was
interrupted by too many waterfalls and gorges as it fell from the central
plateau of the continent to the sea. Navigation inland from the coast
was to prove largely impracticable.

The river rises in the great massif of central Angola and then sweeps
round in a huge curve through western Zambia where in the western
province (formerly Barotseland) it waters the flood plains until
becoming the boundary between Zambia and the Caprivi Strip. Thereafter it passes through a number of points of great importance to the
various countries through which it flows or between which it acts as a
boundary. For the best part of a hundred miles it divides the Caprivi
Strip from Zambia until it reaches its confluence with the Chobe river.
Here, at the Kazungula crossing, is Botswana's only exit from the
region of white domination that surrounds her. From that point eastwards for another 500 miles the river forms the boundary between
Zambia and Rhodesia. Along this stretch are three vital points: the
Victoria Falls where there are both road and rail crossings as well as
power stations; Lake Kariba and the dam which is also a crossing point;
and the Chirundu crossing. A hundred miles farther east from
Chirundu at Feira the river passes into Mozambique, and in the centre
of the Tete province encounters another great gorge — the first inland
from the sea — at Cabora Bassa. The Portuguese have converted this
gorge into the river's second artificial lake: it is the largest hydro-
electric scheme in southern Africa. From the dam to the Indian Ocean

the river is navigable.

The river therefore never provided a major waterway into the interior. In the colonial age it was a lure to explorers like Livingstone as they sought for means to open up the continent, and a convenient boundary to the statesmen of Europe as they worked out on their maps in Berlin or London the disposition of Africa in their various empires. In recent years, however, the river has become a source of power with hydro-electric developments at Victoria Falls, Kàriba and Cabora Bassa, the last two of them also becoming major sources of political contention. The Kariba dam was to be the economic symbol of the ill-fated Central African Federation although the Kafue gorge in Zambia ought to have been developed in preference to Kariba for economic reasons had the Federation been a genuinely non-racial union. Instead, the politics of Salisbury ensured that it was Kariba and that control of it lay to the south of the river away from Northern Rhodesia. The Cabora Bassa dam was to be another experiment in colonial domination, this time by the Portuguese, although with the coming to power of Frelimo in 1975 the dam instead will provide the new African government with a source of income and potential influence that will stand it in good stead.

After 1964, following the break-up of the Central African Federation, the river became the demarcation line between independent black Africa to its north and white-ruled Africa to its south; it also then developed into a highly militarized border and a line of great strategic importance — for South Africa in the Caprivi Strip; for Botswana at the Kazungula crossing; and for Rhodesia and Zambia along their joint border. In the period up to April 1974, when the coup took place in Lisbon, the Portuguese looked to the river as the final defence line where they had to hold Frelimo if its forces were to be prevented from sweeping southwards and dominating the whole of Mozambique (as in fact they were already beginning to do by April 1974 anyway). For the decade 1964—74 the river became a focus of great strategic importance as well as one of the most sensitive boundaries on the continent even if as a highway it did little business.

Of all the countries it touches, Zambia is most affected by the Zambezi. In the west the annual flooding of the Barotse plain is part of the lifestyle of the Lozi people. Along the border with Rhodesia in the years following U.D.I. in 1965, it became a military barrier: there have been many incidents including South African and Rhodesian incursions across it; the row with South Africa in 1970 when she denied that Zambia and Botswana had a common frontier at Kazungula; a number of Zambian deaths as a result of Rhodesian-placed mines; the 1973 shooting of two Canadian girls by Zambian troops with its subsequent international repercussions.

It has also acted as the barrier — heavily guarded from the south by the South African military and the Rhodesians — across which Z A N U and Z A P U guerrillas went to launch and then maintain their activities in Rhodesia. After December 1972 the campaign in north-eastern Rhodesia developed into a guerrilla war which only came to a pause for the 'détente' talks that got underway in late 1974 and continued throughout 1975 and then started again with mounting intensity in 1976.

The river acted as a boundary for Namibia too, providing the demarcation line between the eastern end of the Caprivi Strip and the free north: from 1966 onwards (until détente froze such activities) S W A P O sent their men into the Caprivi Strip to organize resistance against Pretoria's illegal occupation of their territory.

For Rhodesia its significance is twofold: as a boundary and as a source of industrial power. It is the boundary not simply with Zambia but with the whole of independent black Africa to the north, and in this respect it became for Rhodesians a symbol — the line between them and oncoming majority rule. Their national servicemen joined regular soldiers in patrolling it and — before the détente exercise — the South Africans also sent their troops and para-military units to help defend it. Its other value for Rhodesia comes from the power produced at the Victoria Falls and from the huge Kariba dam.

For Mozambique the Zambezi River provided an historical means of access to the interior, at least as far as Tete. During the decades of their bitter war against the Frelimo nationalists the Portuguese came to regard the river as the border south of which the guerrillas must not be allowed to penetrate if Mozambique was to be held: in strategic terms during the years 1970—74 it became increasingly clear that this Portuguese assessment of the river as a crucial line to be held in the war was correct. It was as part of their struggle to hold on to Mozambique that the Portuguese embarked upon the building of the Cabora Bassa dam. The dam was to be a source of power and agricultural develop- ment to enable large-scale white settlement to take place in the area; its lake, it was hoped, would act as an easily controlled barrier to southward movement by Frelimo forces; and its cheap electricity, exported to the huge South African market, would provide the Portuguese with much-needed revenue. Revolution in Portugal, however, brought Frelimo to power in Mozambique, and the new government was faced with the problem of what to do with the power from the dam because South Africa alone has the market for the electricity, but Frelimo's ideology does not easily admit trade with Pretoria.

In 1976 the Frelimo government came to an agreement with the Portuguese to amortize the debt on the dam — which it had accepted —

at a rate of 3% a year over 30 years. The dam is full, the transmission lines to South Africa are ready. Only by selling the power to the Republic can Mozambique obtain the money to repay the Portuguese. The political decision to sell the power to South Africa has been taken in principle. As of June 1976, however, the word to proceed with sales had yet to be given.

For Malawi the Zambezi provides a means of transport to the Indian Ocean, since the Shire river debouches into the lower navigable Zambezi; however, the amount of traffic on this route is hardly great.

The river became so important during the 1960s, after U.D.I. in Rhodesia, that South Africa started to treat the Zambezi as its own forward line of defence against incursions from independent black Africa. In consequence she stationed her troops in the Caprivi Strip and increased the numbers in Rhodesia along the Zambezi. At the same time she provided assistance to the Portuguese for both their wars in Angola and Mozambique. Then came the coup in Lisbon in April 1974. This changed all the strategic considerations of the area.

From South Africa's point of view, as long as the Portuguese held Angola and Mozambique it was essential to secure the weak centre, and this meant securing the line of the Zambezi where it formed the northern border of Rhodesia. Starting in 1961 with the outbreak of fighting in Angola the Portuguese were to build up two armies in their southern African colonies, so that by 1970 they had 60,000 troops in Angola and the same number in Mozambique. The presence of these formidable Portuguese armies in Angola and Mozambique throughout the period 1965 to 1974 persuaded South Africa that it could pursue a policy whose basis was to support the regimes in a cordon of white-controlled territories between the Republic and independent black Africa — which she did throughout this period. The weak link in this cordon was Rhodesia, because where the Portuguese were legitimately present in both their colonies and had two highly sophisticated and efficient (by African standards) European armies to call upon, the Rhodesians had neither of these advantages. U.D.I. was a rebellion which ensured that the spotlight of adverse world criticism was focused upon Rhodesia; Salisbury could never claim legitimacy, and neither the Portuguese nor the South Africans were to recognize the Smith regime despite the help they gave to it.

Although Rhodesia inherited the main armaments of the defunct Central African Federation when it broke up in 1964, she did not have the white manpower for more than a small defence force. The Rhodesian army consisted of 3,500 men; the airforce of a further 1,200 whose 45 mixed planes, however, proved sufficient for counter-insurgency operations. The worst strains upon this force for Rhodesia during the decade after U.D.I. were threefold: the difficulty in obtaining spare parts or replacement equipment as a result of sanctions; the strain upon her white population whose men, to the age of forty-five, were subject to both military service of one year and subsequent recalls of a month at a time which by 1974 could be as many as three in a year; and the adverse effect the guerrilla situation had (especially in the period December 1972 onwards until détente began) upon would-be immigrants. These were urgently needed to reinforce a white population which was growing steadily older as its young people, especially the men liable for national service, left the country. All these difficulties were to grow markedly worse in 1976 once the détente exercise had collapsed and Mozambique had closed its borders with Rhodesia. As a result of this over-stretched defence situation in Rhodesia, South Africa embarked upon a policy of committing troops to the territory in 1967, following a major guerrilla incursion, and by 1969 there were reported to be 2,700 South African troops stationed in Rhodesia.

Following the Sharpeville massacre in 1960 and the subsequent outflow of capital from the country which warned Pretoria of its potential world isolation, the government embarked upon a steady policy of expanding its armed forces. By 1969 the country's forces were formidable in African terms and substantial in world terms especially with regard to the sophistication of arms. In that year the army numbered 16,200 (5,700 regular and 10,500 Citizen Force — South Africa's National Service). The total Citizen Force numbered 55,000 although no more than 10,000 were normally under arms. In addition there were 60,000 Commandos — part-time militia — in 210 units throughout the country. Army equipment consisted of Sherman and Centurion tanks, Panhard armoured cars, Saracen armoured personnel carriers, Cessna reconnaissance aircraft, and FN rifles which are manufactured in the Republic.

There was a large police force, much of it para-military, of 28,000 men, and an airforce of 4,700 men equipped with Sabre MVI fighters, Mirage III fighter bombers, Canberras and Shackletons, Buccaneer light bombers, helicopters and transports. The next year — 1970 — South Africa increased her defence budget and raised the size of the army to 10,000 regulars and 22,300 Citizen Force while the combat craft of the airforce were increased to 230, including 80 helicopters. Further massive expansion of South Africa's forces took place between 1970 and 1974 as the war situation in the two Portuguese colonies worsened — for the Portuguese — so that by the beginning of 1974 South Africa had 17,300 regular forces, 92,000 trained reservists and 75,000 para-military commandos, while her defence budget had increased to R481 million ($716 million). All this was in response to two things: the need as Pretoria saw it for her armed forces to be able to maintain internal security, which is stated – always – to be their first priority; and the need to supply growing support to the 'cordon' countries to her north.

Regular military liaison between the South Africans, Portuguese and Rhodesians had become established by the early 1970s. More important were South African commitments in Rhodesia and Namibia. South African troops and para-military police along the Zambezi and elsewhere in Rhodesia between 1967 and 1975 have been variously estimated at as low a figure as 400 and as high as 5,000; particularly in the early 1970s, their presence along the Zambezi released Rhodesian forces for their campaign against Z A N U in the north-east. In the early 1970s when the South African government would only admit to a figure of 400, both the liberation movements – Z A N U and Z A P U – and the Zambian army estimated the South African presence at between 2,000 and 5,000 men.

As early as 1965 the South Africans were building a large £8 million air-base at Katima Mulilo in the Caprivi Strip: it was the Republic's most forward strike base. There were Beavers, helicopters and Mirage fighters based at Katima Mulilo as well as a battalion of troops, while another battalion was stationed at the end of the Strip near Kazungula. This concentration of South African forces in the Caprivi Strip and along the Zambezi in the late 1960s and early 1970s indicated clearly where Pretoria judged its front line to be. On the other side the Zambian Army was only 4,000 strong and lightly equipped, while its airforce of 300 men had only 4 Dakotas, 4 Caribou Transports, 2 Pembroke Squadrons, 6 Beavers and 2 Chipmunks – a force quite incapable of dealing with any serious confrontation from the south.

This military balance that overwhelmingly favoured the south made the Zambezi a forward defence line of immense importance to all three of the white regimes. It ensured equally that Zambia was only too anxious to move the confrontation southwards, away from the river. Then the April 1974 coup occurred in Lisbon and the situation changed fundamentally overnight.

As long as the Portuguese managed to hold on in Mozambique and Angola it made sense for South Africa to maintain troops on the Zambezi and support Rhodesia so as to complete the cordon of white-controlled countries between herself and independent black Africa. But once the Portuguese had decided to pull out of Africa and, especially, once the forces of Frelimo had effective control of Mozambique as they did by September 1974, South Africa had to reconsider its overall military strategy. The policy of maintaining a cordon between the Republic and independent Africa had collapsed; instead South Africa had to police some 500 miles of border with an independent Mozambique and if there was to be infiltration of guerrillas into South Africa it would be across this border. Since the stated object of maintaining troops on the Zambezi had been to prevent such infiltration there was no longer any strategic point in remaining there (an analogy could be drawn here with France in 1940 holding the Maginot line after the Germans had marched into Belgium). Once Pretoria had accepted the fait accompli of a Frelimo triumph in Mozambique, the result was to attempt détente.

There had been no overt talk of détente until the Portuguese collapse. Thereafter it became manifest (from about June 1974) that South Africa was adopting a new policy with three separate facets. First, a willingness to sacrifice Rhodesia by forcing Smith to come to terms with his nationalists; second, and more reluctantly, a decision to speed up some form of political handover in Namibia which envisaged if necessary the surrender of control of the territory; and third, general détente with black Africa,which, hopefully, would include at least trade relations with newly independent Mozambique. None of this meant a change of heart in Pretoria. It simply meant that the South African government had read the new strategic situation correctly and was changing her policy but not her objective accordingly. Her objective remains the maintenance of white supremacy in the Republic. Her policy in order to further this changed from that of maintaining a cordon between her and independent Africa to one of détente, to buy time. This meant sacrificing the remnants of the cordon — Rhodesia and Namibia — while withdrawing into her historic borders with the Limpopo as her northern defence line.

Thus the Lisbon coup of April 1974 came as a turning-point for southern Africa: before the coup the Zambezi acted as a strategic divide of immense importance; following the coup during the détente exercise it began to look for the first time in many years at least possible that the river would revert to a simpler role as a boundary, an indifferent highway and a source of hydro-electric power.

Had the whole length of the Zambezi been navigable the history of southern Africa might have been very different. Instead, the river is only navigable for some 400 miles from its mouth through Mozambique until interrupted by the Cabora Bassa gorge. It is again navigable above Cabora Bassa and beyond the confluence of the Luangwa up to Kariba. Upstream of the Kariba lake it is once more interrupted by the Victoria Falls and then the falls of Ngonye, before being navigable for light draught barges through the Barotse plain. With the completion of the dam at Cabora Bassa it is possible that new navigation arrangements can be made above and below the dam so as to increase the general use of the river as a highway.

During the nineteenth century the navigation possibilities of the

Zambezi were long in doubt and so a great deal of exploration attention was to centre upon the river's potential. In 1856, while in Tete, Livingstone recorded that he had been told of some small rapids further upstream which he had missed in his earlier explorations and did not then deem worth investigation. They were, in fact, the Cabora Bassa rapids and gorge. As a consequence of this omission on his part Livingstone was to persuade the British Foreign Office of the possibility of opening up the interior by means of the Zambezi, since he worked on the assumption that there would be no major impediment to navigation. When he did finally see the gorge and realized that it made further navigation inland an impossibility, Livingstone saw that his whole plan for exploration and opening up the interior had been shattered; he then turned to the alternative of attempting the same thing through the Shire river. As a consequence, British attention turned sooner than otherwise would have been the case to the Shire and Nyasa highlands.[1]

While HM Consul at Quelimane, Livingstone discovered that the Portuguese had no intention of making the Zambezi an independent waterway. As it happened, the first steam vessel to use the lower Zambezi was the *Ma-Robert*, named after Mrs Livingstone, in 1858. The opening up much later of the railway from Beira to Chindio at the confluence of the Shire and Zambezi diverted much of the traffic that might otherwise have passed down the Zambezi's lower waters, while the later construction of the Tete-Moatize line further deprived the river of traffic.[2] By 1891 when a minor Anglo-Portuguese confrontation took place — in the high noon of the 'Scramble for Africa' — Britain obtained the right of navigation along the Zambezi from Beira into the landlocked British territories, that is, up the Shire into Nyasaland and Zambesia.

During the Scramble for Africa the Zambezi was a focus of high politics. Lord Salisbury — an enigmatic and offhand imperialist — appeared to regard some British acquisitions in Africa as vital, while others he dismissed as though they could have no place at all in imperial strategy. As a writer in the *Fortnightly Review* at the end of the 1880s asserted — although he claimed it was not generally known in Britain — there had been a proposal to cede to Germany a strip of territory extending across the continent from east to west, north of the Zambezi, which would have barred the passage of the iron track which 'must ultimately join the Cape to Cairo and carry civilization through the heart of the Dark Continent'.[3]

Tensions between Britain and Germany had been mounting during the Scramble so that by 1890 both powers were ready for an accommodation. There followed the Anglo-German Treaty of that year which attempted a broad settlement of their outstanding differences in Africa. One part of it was the cession by Britain to Germany of a northern stretch of the Bechuanaland Protectorate, recently established, in order to provide the Germans with access to the Zambezi river from their colony of South West Africa. In theory this was to give them a means of traversing the continent by the river from their colony to the Indian Ocean. In practice – because of the various bars along the river – it meant nothing of the sort. The corridor ceded to Germany was twenty miles wide and some 300 miles long. The Germans had already delimited their boundaries of South West Africa with the Portuguese in 1886; now in 1890 they did the same – to advantage – with Britain:

> During the negotiations with Britain Germany successfully argued in favour of an access to the Zambesi River, apparently envisaging an ultimate link with the German colony in East Africa. This strange claim sealed by bartering with African soil, resulted in the narrow appendix of the Caprivi Zipfel running from the Okavango River to the distant Zambesi and named in honour of the German Chancellor Count von Caprivi.[4]

The Anglo-German Agreement was signed on July 1st, 1890, and included: the handover to Germany by Britain of Heligoland; the demarcation of the two countries' spheres of influence in East Africa, including the relinquishing to Britain by Germany of its claims to Zanzibar and the coastal strip; and the giving to Germany by Britain of the Caprivi Strip so that through her colony of South West Africa she should have free access to the Zambezi. The Strip could be seen as a part of Germany's effort to push her advance eastwards across the whole continent, but this ambition was thwarted by the British advance northwards from the Cape and – also occurring in 1890 – Rhodes's occupation of Matabeleland. Even so, as one commentator put it:

> In South-West Africa he [Lord Salisbury] was not so successful, for yielding to the desire of the Germans for some means of access to the upper waters of the Zambesi, he allowed them to occupy a strip of territory nearly three hundred miles in length, stretching from the northern boundary of the colony to the junction of the rivers Chobe and Zambesi. This wedge of territory, generally known as the 'Caprivi Strip', was in reality intended to provide a wedge against British expansion in Barotseland, and to prevent direct communication through the Bechuanaland protectorate with the Portuguese colony of Angola, because German access to the Zambesi at any point above the Victoria Falls could be of small commercial value.[5]

This indeed was, and still is, true, emphasizing what has always been a major aspect of the Zambezi's importance: its strategic value. When the new Chartered Company of Rhodes protested at the yielding of the Strip to the Germans, Salisbury commented sarcastically: 'It is . . . over an impracticable country, and leading only into the Portuguese possessions, into which, as far as I know, during the last 300 years there has been no very eager or impetuous torrent of trade. I think that the constant study of maps is apt to disturb men's reasoning powers.'[6]

The Strip, which proved politically and strategically vital in the 1960s and 1970s, was not, in fact, occupied at all by the Germans after its acquisition until 1908 when a post was established there. The Anglo-German Agreement of 1890 certainly settled a number of outstanding points between the two powers and, despite the Rhodesian protest at the British creation of the Strip for the Germans, by 1965 they must have been pleased at its existence since the alternative would have been a long Botswana border with Angola and Zambia. At the time Salisbury described the whole Agreement, whereby the principal British concession was then seen as the little island of Heligoland in Europe and her principal gain as the Zanzibar coastal strip of East Africa, by saying: 'We have exchanged a trouser button for a suit of clothes.'

The Strip changed hands when German South West Africa became a League Mandate under South Africa in 1919. In modern times the Republic has come to regard it — through the 1960s and 1970s — as her northern frontier with black Africa. It has been divided into two projected bantustans — Eastern Caprivi and Kavango — while the South African government has invested £1.3 million in the Vungu Vungu agricultural project in an attempt to woo the local people along the Okavango river. From 1960 onwards the Strip became a military zone as far as South Africa was concerned, in direct contravention of the terms of the Mandate, and early in the 1960s South Africa built the huge Katima Mulilo air-base which is capable of handling the largest jets. She also has military garrisons in the Strip and since 1966 they have had to face increasing guerrilla activity from S W A P O, the South West African People's Organization. From 1970 onwards with a war in neighbouring Angola and an escalating guerrilla campaign in Rhodesia, several thousand South African troops have, from time to time, been stationed in the Strip, and helicopter and other patrols have been carried out from it along the Zambezi, sometimes leading to the violation of both Zambian air-space and territory. In March 1973 the South African Government allocated R56 million for the building of an all-weather road (for military purposes) from Grootfontein to Katima Mulilo.

The Strip has always been strategic: it was invented for strategic reasons – that is, to satisfy German demands in an area where Britain was prepared to make sacrifices to safeguard what she considered to be more important objectives elsewhere on the continent. According to the terms of the League Mandate South Africa may not have military forces in the Strip; not only has she had numerous exercises there but she has established bases there as well, while at the start of the 1960s she tried to persuade Britain to set up a school of tropical warfare in Caprivi.

Again and again over the last century the Zambezi has been regarded in strategic terms. In 1888 the Glasgow Chamber of Commerce expressed their ' . . . sense of the importance of extending British influence and administration in transcontinental territories in South Africa up to the Zambesi and of securing its free navigation as an international river.'[7] This was during the days of the Scramble. In the 1950s, during the brief experiment of the ill-fated Central African Federation the Kariba dam was built as a source of power for the Federation. Yet strategic considerations clearly played a major part in this, so that Welensky insisted upon the Kariba development instead of the alternative Kafue one since even then, it seemed, the whites in Salisbury did not trust the political durability of their much vaunted political experiment and wanted, in the event of a break-up, to be in control of the power. The Kariba lake later provided a convenient physical barrier to guerrilla infiltration across the border from Zambia into Rhodesia.

The river thus became the political focal point between independent black Africa and white minority-ruled southern Africa. For the decade following U.D.I. in Rhodesia it has been patrolled by power boats, watched by troops and police on land, and helicopters from the air. Its banks have been mined and used at a number of points by freedom fighters crossing southwards to fight in Rhodesia, Namibia, Mozambique, sometimes going on to South Africa. From 1967 onwards South African troops and para-military units were stationed in Rhodesia – possibly up to 5,000 at times – and most of these were along the Zambezi. At the time of U.D.I. the fear was expressed in Zambia that the Rhodesians would attempt to blow up Kariba; Zambia then went ahead with the development of the Kafue gorge hydro-electric scheme so that she need not be dependent upon Rhodesia for power. She followed this by tackling the development of the north bank power station at Kariba for the same reason. In 1970 South Africa denied the existence of any border between Zambia and Botswana at Kazungula – although later she climbed down – as part of her offensive against black Africa and because she then regarded the Zambezi as her strategic frontier.

The crossing at Kazungula in the early 1970s presented a curious picture against the background of the known border tensions of the area: the little one-car ferry would chug across the river from Botswana to Zambia in answer to a summons upon an iron triangle. South African soldiers among the reeds at the tip of the Caprivi Strip would peer through binoculars to take the number of any car crossing to Botswana; white Rhodesian soldiers behind barbed wire fifty yards from the landing stage in Botswana would eye the traveller curiously; and nine miles away at the Chobe River Hotel in the game park white South African and Rhodesian soldiers in mufti would relax to enjoy the pleasures offered by a black-ruled country. By 1973 both sides of the river between Zambia and Rhodesia had been mined: by the middle of the year mines planted from the Rhodesian side had accounted for twenty Zambian dead and many more injured; then an international outcry followed the shooting of two Canadian girls by Zambian troops at the base of the Victoria Falls.

The Zambezi in its many manifestations is also a river of industrial power. There are three hydro-electric developments along its length: the Victoria Falls, Kariba and Cabora Bassa. Kariba was the great development of the Central African Federation to provide power to lubricate the economy of a white political dream; it became instead the symbol of white oppression for Northern Rhodesia and Nyasaland. Then, in 1968, Portugal embarked upon the building of the Cabora Bassa dam. By 1973 the first construction phase had been completed. The objectives behind the dam were to transform the territory's economy by settling one million Portuguese in the area; to link the economy more closely with that of South Africa; and to associate European and American business interests with the maintenance of Portuguese control. Portugal almost pulled off these objectives before her 1974 change of government. The *Johannesburg Star* had commented as early as 1969:

> What is today desolate and almost unpopulated country will
> provide a far better living for hundreds of thousands of people
> and their presence and their prosperity will, there is every
> reason to believe, not only make it easier for the Portuguese to
> take their own people with them in resistance to terrorism, but
> also provide protection for others in Southern Africa. Cabora, in
> other words, makes a great deal of political as well as economic
> sense.[8]

By the start of the 1970s Frelimo instructions to their militants south of Zambezi were: (a) to intensify the mobilization of the people; (b) to consolidate Frelimo political and military structures; and (c) to begin guerrilla operations there. By early 1974 it was clear that these instructions had been well carried out. The Cabora Bassa dam was the largest hydro-electric scheme in southern Africa, and since Mozambique

Rhodesia came into being as a result of the expansionist policies of Rhodes. He was then fighting for political and economic control in the whole of southern Africa so that Rhodesia was a part of his outflanking operation directed against Kruger in the Transvaal. A landlocked state, Rhodesia needed its own outlets to the sea. In 1892 the B.S.A. Company began the construction of the railway inland from Beira across Mozambique to Umtali which it reached in 1898; it reached Salisbury a year later in 1899. By 1902 the Bulawayo-Salisbury line was linked into the South African system.

Rhodes, however, never forgot his dream of an imperial railway from the Cape to Cairo. The first major section of this from the Cape through Bechuanaland reached Bulawayo in 1897, skirting the hostile Transvaal. By 1905 it had reached Northern Rhodesia and by 1910 the Copperbelt. Much later, between the wars, the line from Northern Rhodesia was pushed farther north into the Congo to the new copper mines at Katanga: it arrived at the rich mining area of the Congo in time to forestall the attractions of the Benguela railway and so enticed southwards the copper freights of the region.

Rhodesia Railways became the transport backbone of both the Rhodesias: the railway's most important section runs 1,270 miles from the Mozambique border through Rhodesia and then Zambia to the border of the Congo (Zaire) north of the Copperbelt. By the 1950s, however, this Beira line had become heavily congested and a new line to relieve the pressure was opened from Gwelo to Lourenco Marques. The system was linked into that of South Africa by the 398-mile stretch of railway through Botswana which handles mainly Rhodesia-South Africa traffic.

As it is, Rhodesia railways are one of the most heavily used in Africa: before U.D.I. they included 2,708 miles of track − 1,356 in Rhodesia (excluding the new Rutenga to Beit Bridge line completed in 1974), 643 miles in Zambia (separated off after U.D.I.), the 398 miles of track through Botswana that by 1975 the government of Seretse Khama was contemplating nationalizing, 112 miles in South Africa and 198 miles in Mozambique.

Rhodesia has three railway exits to the sea: the Cape Town-Ilebo (Zaire) railway cuts through Rhodesia, forming a part of Rhodesia Railways while crossing the country between Plumtree and Livingstone; the line from Salisbury to Beira; and the line from Gwelo to Lourenco

already possessed six other hydro-electric projects covering all i
the purpose of the dam was clearly political; it was to provide c
power for South Africa. Of the projected 2,200mw of power So
Africa agreed to purchase an initial 1,000mw a year.

As a highway the Zambezi has been largely a failure. As a str
divide it has few equals in Africa or elsewhere. Over the last hui
years it has figured prominently in much of the history of soutl
Africa: as a magnet for explorers in the pre-Scramble era; as a
strategic attraction in the imperial age; as the dividing line after
decolonization when the confrontation between black Africa ai
white-dominated south became a fact of African politics; and ii
1974—5 during the 'détente' era, as the symbolic line across wh
side viewed the other with suspicion. In all these respects it is o
the greatest of African waterways, its history bound up with th
conflicts of the area.

Notes

1. See Judith Listowel, *The Other Livingstone* (Julian Friedmann, Lon
   pp. 156—76.
2. Lord Hailey, *An African Survey* (O.U.P., London, 1957: revised 19!
   p. 1540.
3. Quoted in R.W. Bixler, *Anglo-German Imperialism in South Africa i
   1900* (Warwick & York, 1932), pp. 55—6.
4. J.P. van S. Bruwer, *South West Africa: The Disputed Land* (Nasiona
   Boekhandel Bpk, Cape Town, 1966), p. 73.
5. Evans Lewin, *The Germans and Africa* (Cassell, London, 1915), p. ?
6. Robinson, R., and J. Gallagher (with Denny, Alice), *Africa and the
   Victorians* (Macmillan, London, 1961), p. 296.
7. Bixler, *op. cit.*, p. 55.
8. *Johannesburg Star* (Weekly Airmail Edition), September 6th, 1969.

Marques. The first of these alternatives — northwards through Zambia and Zaire — is theoretical rather than practical and has not been used since U.D.I. Then in 1974 a fourth alternative was added when the new Rutenga to Beit Bridge line was opened, giving a second means of access to the South African system.

The Rhodesian road system is well developed: there are 18,750 miles of roads of which 8,000 are of international importance and ensure communications with Zambia and Mozambique, Botswana and South Africa.

The communications system of Rhodesia is inextricably bound up with her politics. U.D.I. in 1965 changed the existing pattern of communications dramatically. However, it has first to be understood that U.D.I. was only possible in the light of Salisbury's knowledge that its two most vital lines to the sea would remain open: the Portuguese, fighting their own war against Frelimo in Mozambique, had no intention of cutting off Rhodesia or depriving themselves of the revenues derived from her freights to either Beira or Lourenco Marques; and the South Africans, despite their unwillingness to recognize an illegal regime in Salisbury, their annoyance that U.D.I. had been declared at all and their own well-developed partiality for strictly legal behaviour, nonetheless gave Rhodesia all the transport facilities through the Republic that she required — at a price. They did so for two reasons: the emotional support for the beleaguered and defiant whites in Rhodesia from South Africa's own whites; and South Africa's determination to demonstrate — as a long-term safeguard for her own interests — that sanctions would not work.

Believing that communications through Mozambique and South Africa were safe, the Smith regime could afford to declare U.D.I. in 1965. Even so, in the years that followed, the pressures were to mount upon Rhodesia's communications system as first sanctions and then the guerrilla war narrowed Salisbury's options. Sanctions put a steadily increasing strain upon the railways as rolling stock and spare parts could not be replaced and maintenance became ever more difficult. Frelimo successes in Mozambique were threatening to close the railway to Beira when the coup in Lisbon altered the entire position.

With the Portuguese collapse in Mozambique and the possibility from mid-1974 onwards that a Frelimo government in Maputo would close the railways from Beira to Salisbury and Lourenco Marques to Gwelo, the Rhodesian position was potentially extreme. Anticipating that her lines through Mozambique might one day be closed, Rhodesia had examined various possibilities that would give her a direct link into the South African system, as opposed to using the long stretch of railway that passed through Botswana. Following the Lisbon coup and after agreement between Salisbury and Pretoria, a crash programme was initiated and a new line built from Rutenga to Beit Bridge so giving Rhodesia its first direct railway link with South Africa. The line was opened in September 1974. Even so, the Rhodesian Government began at once to examine the possibility of a second direct link from West Nicholson to Beit Bridge. A major problem however was not one of lines at all but of locomotives and potential congestion at South Africa's ports. South Africa is short of locomotives and by 1975 was saying that any increase of Rhodesian traffic that she had to handle would have to be carried by Rhodesian locomotives. The Botswana line is already operating to capacity so that any switch of Rhodesian traffic from the lines through Mozambique would have to be carried on the new Rutenga-Beit Bridge line.

At the end of 1975 South Africa's expanding port facilities could probably have handled all Rhodesia's exports and imports — if it had become essential to do so; yet South Africa itself is heavily dependent upon using its own line to Lourenco Marques (now Maputo). Thus the communications of both countries waited upon political decisions in Lourenco Marques. In March 1976 Mozambique closed its borders with Rhodesia, so making Salisbury entirely dependent upon communications through South Africa, commensurately narrowing the options open to Rhodesia. As a consequence Rhodesia faced a situation in which South Africa would only handle her freight after all its own requirements had been met — and Rhodesia had no other alternatives.

Meanwhile Rhodesia's policies over the years following U.D.I. had produced more than one confrontation with her northern neighbour, Zambia, so that the dividing line between them, the Zambezi, had developed into a frontier of permanent tensions. For Rhodesia, faced with guerrilla threats from the north, the Zambezi did at least provide a relatively good border to defend and patrol. Far more important in the relations between the two countries was the fact that following U.D.I. Rhodesia held a major card in all her dealings with Zambia because the latter's communications were then almost entirely routed through her neighbour to the south (see Chapter 6, Zambia). Moreover, once Rhodesia was certain that Britain was not going to take military action against her, she proceeded to use her control of Zambia's communications as a lever and bargaining counter in the years that followed, since Zambia's copper had to use the route through Rhodesia and out to Beira; it was not until 1973 that this position was to change. Then, following the rapid worsening of the guerrilla war in the north-east of Rhodesia, in a mistaken move — mistaken in the miscalculation as to its outcome — Smith closed the border with Zambia on January 9th, 1973.

He did this in an attempt to force Kaunda to clamp down upon the activities of the liberation movements, assuming that Zambia would be brought to its knees economically if it could not quickly re-open the border and use the railway to get its copper out to Beira. Smith, however, had underestimated two things: the extent to which by then new routes to the north had been developed by Zambia and could quickly be brought into use as alternatives; and the degree of international assistance that Zambia could call upon. The result was that Zambia diverted the balance of her copper exports (those that had been passing through Rhodesia to Beira) northwards through Tanzania or westwards through Angola and as a result Rhodesia was to lose more than £500,000 a month in Zambian freight payments – currency that she dearly needed. This loss demonstrated how much the success of Rhodesia Railways depended upon handling non-Rhodesian freight. The closure of the border, however, meant that Rhodesia was narrowing down her own options, as she was soon to realize. Her dependence upon her two white neighbours – Mozambique and South Africa – was becoming greater, while her room for manoeuvre was growing commensurately less.

In January 1973 the Portuguese still controlled Mozambique although their troops were fighting a losing battle in its northern provinces. By January 1974, however, ominously for Rhodesia, Frelimo forces had begun to attack the Beira line, something they could have done earlier but had held off, at least in part, in order not to embarrass Zambia. By closing the border and stopping Zambia from using the railway to Beira Rhodesia had released Frelimo from such inhibitions. Thereafter the pressures mounted upon Rhodesia: Frelimo was committed to applying U.N. sanctions after independence which meant closing the two rail routes to Beira and Maputo; while the 800 miles of rough country along the border between Mozambique and Rhodesia – if opened to Z A N U guerrillas – would present the Rhodesian security forces with far greater headaches than the northern Zambezi frontier. Then the détente exercise, mounted by Kaunda in October 1974, gave Rhodesia a respite.

The chances of détente working in Rhodesia – that is, the white minority actually sharing and then handing over power peaceably to the black majority – were minimal, when the diplomatic exercise started. The exercise collapsed in early 1976 and then Mozambique closed her borders to Rhodesia so forcing it to rely entirely upon South African communications. Quite apart from South African determination to force the Salisbury government to come to terms with its nationalists (for Pretoria's own selfish reasons) the possibility of being entirely dependent upon transport through the Republic was not an appealing one in Salisbury at the beginning of 1976. In the first place the cost of exporting through South Africa is higher than through Mozambique. In addition the South African ports are heavily used and becoming congested, and South African Railways is also under increasing strain. On the other hand the reopening of the Suez Canal should ease the congestion on South Africa's ports, and the new port of Richards Bay went into operation at the end of 1975.

Indeed, at the beginning of 1976, Rhodesia as much as her two northern neighbours, Zambia and Malawi, found herself a prisoner of her colonial heritage: this was so both in terms of the communications network available to her and in terms of the political and racial problems she had inherited from the imperial age. During the colonial era Rhodesia had been the nub of the railway system that served Northern Rhodesia (Zambia) and stretched as far as the Congo (Zaire) as well as Botswana and Mozambique. After U.D.I. she assumed, and for long was able to get away with the assumption, that because of the way the railway system had developed Zambia had still to use it and so, in a sense, was her economic prisoner. By 1975, however, with the coming of the T A N Z A M railway and other developments Zambia — despite continuing difficulties — was fast making a bid to become a new rail centre in her own right; a position that she may well maintain in the future even when majority rule eventually comes to Rhodesia.

Just as circumstances in the period 1965—75 forced Zambia to look for new routes and widen her options, so too did circumstances over the same period of time inexorably narrow down the options open to Rhodesia, until by the end of 1975 she was for all practical purposes entirely dependent upon the goodwill of South Africa.

It was because his options had been narrowed down and not for any other reason that Smith embarked upon the détente exercise of 1974—75. He and his government did so reluctantly and they adopted familiar tactics of constant stalling, apparent willingness to talk and subsequent drawing back whenever there was even an appearance of a modest agreement being reached. They found increasing difficulty in playing this particular political role, for the simple reason that South Africa was determined to make détente — in Rhodesia — work. When Smith attempted to assert (in an interview on British television in September 1975) a limited independence by ascribing the failure of his talks with the A.N.C. to outside interference he was called to Pretoria and there made a humiliating public apology, since by then his communications had been reduced to those through South Africa and effectively he had no option left at all.

In 1965 Rhodesia dominated a network of communications which she could control to the detriment of her northern neighbour, Zambia. By 1975 major changes had occurred in the fortunes of all the countries in central Africa, and Zambia — although that year was an economic disaster for her — was fast becoming the centre of a totally new communications network that may well establish the pattern for the area for the balance of the century.

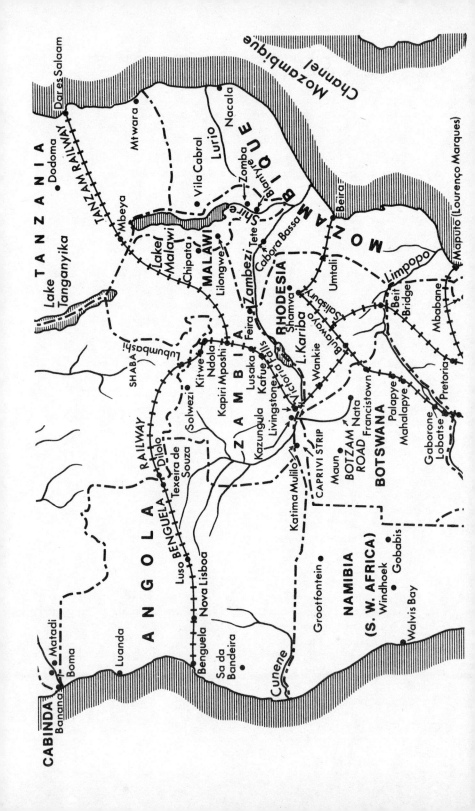

7 ZAMBIA

Zambia is landlocked, sharing borders with eight countries, so that she is absolutely dependent upon good relations with at least some neighbours for transport links to the outside world. Moreover, her major export, copper, is both bulky and heavy and between 50,000 and 60,000 tonnes a month has to be exported to sustain the development and rate of growth that Zambia has attempted to maintain since independence. The copper industry also requires a number of heavy imports — machinery, equipment and mining building materials — so that there is also a bulky volume of imports to bring into the country. As a result of these copper-dominated import-export needs, Zambia is especially vulnerable to sudden changes in the pattern of available communications routes and so far none available to her has been able to provide — for a variety of political reasons — long-term stable services.

On the whole Zambia's rivers are not navigable — over three quarters of the country drains into the Zambezi — and certainly not navigable to the sea. Of Zambia's eight neighbours, Zaire lies to the north, Tanzania north-east, Malawi to the east, Mozambique south-east, Rhodesia south; Zambia touches Botswana briefly at the Kazungula crossing in the south and then for 100 miles it borders the Caprivi Strip of Namibia before its long west and north-west border with Angola. A pedicle strip, the heritage of colonial boundary haggling, has given the Republic an odd appearance, like a 'lop-sided butterfly'.

In the years following her independence in 1964 Zambia became obsessed with transport problems. Her transport system in 1964 was under-developed and almost entirely dependent upon communications through countries to the south. Despite this, Zambia has potential or actual access along more routes outwards than any other landlocked country on the continent. The Copperbelt — the source of her wealth — is 1,250 miles in any direction from an ocean port except to Lobito at 800 miles. There are three basic routes out: to Beira, to Lobito, and to Dar es Salaam. There are also secondary terminal ports for her at Mombasa, Nacala, Maputo, Durban, Matadi and Mtwara.

The history of the country's transport strategy falls neatly into phases. First, there is the colonial era when the only route led to the south; then came the brief initial stage of independence when a start was made in changing this over-dependence upon the south; but following almost at once came the post-Rhodesian U.D.I. period when in order to comply with both U.N. sanctions against Rhodesia as well as her own ideological inclinations, Zambia had to struggle to get her exports out and imports in, if possible, without recourse to the traditional and only established routes through the south. At first a compromise period was to follow when Zambia acknowledged that the southern route through Rhodesia was essential to her — at least until such time as the new rail link to Dar es Salaam was ready; then scheduled for the mid-1970s. Finally came the phase after Rhodesia had closed the border with Zambia on January 9th, 1973, when Zambia began her massive re-routing operation and stopped using the southern route through Rhodesia altogether. Rhodesia had clearly intended that the border closure should be a short-lived affair until such time as Zambia controlled the Zimbabwe guerrillas. It did not work out as Salisbury had planned. When the border was closed on January 9th, 1973, Rhodesia excluded Zambian copper from her ban. Zambia retaliated, however, by retorting that copper also would no longer be sent on Rhodesia Railways and when on February 4th, three weeks later, Rhodesia re-opened the border, it was to receive the reply that as far as Zambia was concerned the border would remain closed. Both South Africa and Portugal were affected by this border closure; Rhodesia had not consulted them.

Over the ten-year period 1964–74 the change for Zambia was to be remarkable. From almost total dependence upon transport through the south she diversified through Tanzania, Zaire, Angola, Malawi and even as far north as Kenya. As a result of U.D.I. and her acceptance of general O.A.U. policies towards the south Zambia had to reroute. There was congestion at Dar es Salaam to be coped with; she met an all-out Malawi effort to attract her business to use the Nacala route once it was opened in 1970; and she found that the Benguela line through Angola was turning out to be the cheapest and most efficient available to her. Thus, as the T A N Z A M neared completion, the effectiveness of the new link was threatened because of the pressures upon the harbour at Dar es Salaam. By the mid-1970s, however, Zambia's alternative lines of communications added up to a substantial number of viable routes.

The T A N Z A M railway will be capable, when fully operational, of handling 1.7 million tons of freight a year both ways, the chief limitation upon it being likely to arise from inadequate port facilities at the railhead at Dar es Salaam. The Benguela Railway to Lobito on the Atlantic coast of Angola is able to handle 60,000 tons of copper a month at more favourable rates than Dar es Salaam. Thirdly, Zambia has the possibility of developing the Nacala route through Malawi and Mozambique — the shortest distance for her to the sea — which is only six hundred miles. President Banda has argued strongly — for his own reasons – that Zambia should not become over-reliant upon only one route. All these routes are or should be effective in 1976 once Zambia came to terms with the new M.P.L.A. government in Angola, so ensuring that she could again use the Benguela railway to Lobito – when the tracks damaged in the recent fighting had been repaired, while the southern alternatives through Rhodesia and Beira were cut as of January 1973. What has to be realized about the changes Zambia faced in the ten-year period following her independence is that they were the result of politics rather than economics – and the lesson becomes increasingly clear: that it is not possible to talk of economic development of communications in isolation from their political significance.

In 1964 Zambia was entirely dependent upon a transport supply system centred primarily upon Rhodesia and thereafter on South Africa and the Portuguese colony of Mozambique. At that time 38 per cent of her imports came from Rhodesia and 22 per cent from South Africa. Copper exports vital to Zambia's economy were handled by Rhodesia Railways, coal came from Wankie in Rhodesia, power was controlled at Kariba in Rhodesia and oil came in through Rhodesia as well. Zambia is one of the three biggest copper exporting countries in the world (the other two are Zaire and Chile) and it is to transport this heavy material to foreign markets and to import the heavy input materials needed to maintain the industry (developed since the early 1920s) that Zambia's transport links are so vital to her. Since her independence Zambia has made tremendous strides in her overall development: the social services, industrialization, modernization. To do so Zambia had to obtain large quantities of capital goods. Demand for consumer goods also increased.

To give an idea of the impact of independence it should be recorded that the First National Development Plan (which followed a transitional plan) provided for an increase of government investment of 100 per cent for the years 1966–70 over the rate of investment envisaged in the 1965–66 period of the Transitional Plan – a total capital outlay of £429 million of which the public sector was to provide £282 million. By the time the Second National Development Plan was launched for the period 1972–76, total investment was then expected to be K2,000 million (approximately £1,500 million), a jump of more than £1,000 million over its predecessor. The Transitional Plan had already referred to 'insufficient railway capacity'. The First National Development Plan states plainly that 'Zambia cannot freely develop her economic potential unless she can import and export goods, without restriction from outside.'[1]

Both plans take note of another essential ingredient of Zambia's development requirements: the need to diversify its transport links as a matter of politics rather than economics. Both in geographical and in political terms, the country has found itself in the forefront of the

69

independent African states that have had to face those dominated by white minority regimes since the beginning of the independence era. Yet, economically, Zambia was developed as part of the southern region and was regarded from Pretoria as an integral part of her economic sphere of influence. Zambia's non-racial policies, however, inevitably brought her into conflict with South Africa and, after U.D.I., with Rhodesia, and these conflicts sharpened the need for a diversification of communications links. Of her eight neighbours no less than four were hostile to Zambia before the April 1974 coup in Lisbon; yet these were the four — Rhodesia, South Africa, Angola and Mozambique — upon whose goodwill (or at least willingness to co-operate) Zambia had to depend most for transport in and out for her copper and the vital imports needed for her economic growth. At independence copper was exported along two main routes: the railway through Rhodesia and thence to the ports of Beira and Maputo in Mozambique; and the Benguela railway through Zaire and Angola to the Atlantic. The main route lay through Rhodesia either to the Mozambique ports or to South Africa.

It is difficult for a country placed as was Zambia at independence to achieve autonomy. Yet autonomy had to be pushed to the maximum and communications were the key to this. Fortunately Zambia had both a government with the determination and the financial means (from copper) to do so. The copper industry provides 95 per cent of foreign exchange, between 40 and 50 per cent of government revenue, and is a major employer setting the pace of industrialization. Gross Domestic Product rose rapidly following independence. 1963 saw the demise of the Central African Federation in which Zambia, as Northern Rhodesia, had been a partner for ten years with Nyasaland (Malawi since 1964) and Southern Rhodesia (self-styled Rhodesia since 1965). The new Zambia gained control of copper taxation. Indeed, only a few hours before the independence flag was hoisted in October 1964 did the country come to an arrangement with the British South Africa Company (Chartered). This meant that the new government took over the copper royalties. These, together with rising copper prices at that time, gave to the government a quadrupled revenue over the first four years of its independence, providing funds that no colonial government had ever had at its disposal. As a result there was a huge push into development projects over a comparatively short time; the result was a vastly increased volume of import requirements putting added strains upon the already overworked transport system.

In 1973 Smith closed the border between Rhodesia and Zambia 'to teach Zambia a lesson, because of its support for terrorists . . . until Zambia gives an assurance that terrorists will no longer operate against Rhodesia from her territory.'[2] United Nations teams calculated that (in 1972) about 1.4 million metric tons of goods entered the country. While this figure of weight is important, that of volume has been impossible to define: thus, a truck load of consumer goods, while comparable in weight to machinery, cannot be compared in terms of volume. Building materials, giant cranes or compressors for the mining industry have special needs: previously these latter machines had come by road rather than rail, trundling up from the supplier country, South Africa, across the Zambezi bridge at Chirundu. Soon after January 9th, 1973, this narrow, winding road emptied of its ceaseless traffic of heavy-duty vehicles. Instead, only official cars, a few curious sightseers and army lorries travelled the route, together with holiday-makers bound for the Zambian end of Kariba. Even this road traffic had been subjected to pressures by the Rhodesians: in 1970 Rhodesia had imposed a tax on through-traffic partly, no doubt, to hit at what Salisbury regarded as Zambian defiance in using the roads as opposed to relying upon Rhodesia Railways. This tax was one of the tactics Rhodesia employed to try to force Zambia to use the railway, and she advanced the rationale that the trucks were causing serious damage to the roads. This reason, ironically, was to be used by Tanzania also in 1974 when she imposed a ban on Zambian through traffic to Kenya; in her case it was perhaps more understandable given the vast increase in the volume of traffic along that northern route following Smith's border closure to the south and the fact that some of the Tanzanian roads were only secondary ones. Blackmail and control exercised through mastery of the communications system of the area were Rhodesian tactics.

Zambia was colonized not from overseas but from the south. The main impetus had come from Rhodes and his Chartered Company: he wanted the high plateaux as part of his company empire, to gain the copper deposits of Katanga and the hoped-for gold between the Limpopo and Zambezi. The aim of the British at the time was simply to keep the Boers, Germans and Portuguese from further expanding their territories and forming alliances which might be hostile to British interests. Thus Rhodes gained his Charter in 1889 and this allowed him to 'persuade' African chiefs to transfer concessions to him. Rhodes failed in Katanga: there King Leopold II of the Belgians was every bit as acquisitive as Rhodes and was determined to keep the British out of the Congo. The Chartered Company did, however, obtain firm control of Northern Rhodesia — which it did nothing to develop — yet this remained important terrain for Rhodes's projected Cape to Cairo railway. By 1909 the railway had reached as far north in Northern Rhodesia as the Katanga border, having been diverted from its original route because of the mineral deposits discovered at Broken Hill (now Kabwe) in Zambia.

The Katanga mines exported their copper south to Beira. In return the railway carried coal and coke from Wankie in Southern Rhodesia and maize and beef from the white farmers along the line of rail in Northern Rhodesia. The Company initially made little profit from the administration of Northern Rhodesia and only finally established its authority in 1910. In 1924 administrative responsibility was taken over by the British Colonial Office. Locally a Legislative Council, in which virtually only white settlers had the right of representation, was set up. Chartered's 'line of rail' in Northern Rhodesia was a continuation of the line from the Cape across Bechuanaland (Botswana), described by Rhodes as 'the Suez Canal to the North', through Southern Rhodesia from where it crossed the Victoria Falls into Northern Rhodesia at Livingstone.

Much of the colonial pattern remains. Names have changed, so have policies, but the facts remain as before, sometimes with great political inconvenience. Thus Katanga has become Shaba and its copper mines no longer belong to Union Minière but to the state owned Gecamines; yet the copper mines still import coal from Wankie, still send part of their output to Beira and still buy food including maize from Rhodesia. And all this in the years following U.D.I., and despite sanctions. Indeed, this traffic to Zaire increased heavily after the 1973 border closure, almost as if real capacity to Lobito was being cleared for Zambian traffic while Rhodesia Railways took an increasing load from Zaire. Thus, during the first part of 1973 imports for Zaire through Rhodesia were running at an annual average of 19,000 tons, rising to an average figure of 27,500 tons for the same period in 1974: most of this was coke, coal and maize originating in Rhodesia. Exports using Rhodesia Railways averaged 3,000 tons in 1973 rising to 4,000 tons in 1974; these represented mainly copper but also included a certain amount of zinc. Zambia in fact found herself the centre of a highly complicated series of manoeuvres over transport.

Early enough Zambia realized the importance of breaking her dependence upon one main communications link, particularly as that was through the south, for both economic and political reasons. The Transitional Development Plan at the time of independence referred to the possibility of extending the railway network in general and the proposed north-east link with Tanzania in particular.

At that stage the system — Rhodesia Railways — was still jointly owned by the Zambian and Rhodesian governments. Zambia had only a partial influence on the investment programme. Rhodesia's illegal declaration of independence on November 11th, 1965, was more than a political shock for Zambia, for it disrupted the country's economy, upset all the targets originally set under the development programme, and made the re-orientation of transport routes the country's first priority. At the time the only existing alternative railway route was that of the Benguela railway, which passed through the Congo (Zaire) and on through Angola to Lobito. Increased capacity was not easily available. Katanga copper and import requirements had the first call on the Benguela line and the proposed Zambian link with Tanzania was viewed against this background. Politically, the Tanzanian link was also seen in Lusaka as being more desirable because of the Portuguese dominance at that time in Angola.

The First National Development Plan, drawn up before U.D.I. but after the decision to build the Tanzania rail link (the full feasibility study for which was expected at the end of 1966), showed how the colonial pattern had precluded trade links with neighbouring countries other than Rhodesia. The plan proposed the establishment of a railway workshop to meet the needs of existing Zambian rolling stock as well as the additional needs to be met when the Tanzanian link had been built. In the event, however, Zambia came to have two workshops, one at Kabwe built for Zambia Railways, the other at Mpika (completed in 1976), built by the Chinese as part of the T A N Z A M railway. This first plan provided for the purchase of rolling stock (diesels). It also devoted a good deal of attention to roads. Two major roads, the Great North Road to Dar es Salaam, and the Great East Road to Malawi (connecting with the Malawi railway system and so giving access to Beira) were to be developed and fully tarred. The Great North Road approximately followed the line of the new rail but this was desirable: in the case of East African Railways the route from Mombasa to Kampala also ran parallel to a first-class road which acted as a feeder for the railway. A similar development was envisaged in the case of the Great North Road and what came to be known as T A N Z A M – the Tanzanian-Zambian railway.

The plan had hardly been drawn up, however, before all this theorizing about transport strategies in Zambia became obsolete: the road was needed instantly after U.D.I. It was to be labelled the 'Hell Run' as heavy-duty vehicles ploughed up the 938 miles of untarred surface between Dar es Salaam and Kapiri Mposhi on the old Zambian line of rail. One of the immediate effects of sanctions was, in fact, to do more damage to the Zambian than to the Rhodesian economy. The two countries' affairs had been deeply intertwined up to this stage of their histories. It was not for nothing that Zambia had been colonized from the south: not only were Zambia's transport routes all oriented southwards; its supplies – consumer and capital alike – were also at that time largely obtained from either Rhodesia or South Africa. During the days of the Central African Federation Northern Rhodesia had become the market for the industries of Southern Rhodesia. Strategic installations such as the hydro-electric power scheme at Kariba were placed in Southern Rhodesia rather than Northern Rhodesia; the pipeline built to supply crude oil to a refinery supplying both territories' needs went from Beira in Mozambique to Umtali in Southern Rhodesia. Thus Zambia was obliged to introduce a carefully manipulated petrol rationing policy – carried out with the help of multinational companies; while Rhodesians were less restricted with regard to fuel because it originally came freely over Beit Bridge from South Africa and later through Lourenco Marques. In the 1970s a new oil pipeline was built in Rhodesia from the Samabula rail junction in the Midlands to Salisbury so as to take the load off the railways. In the meantime, as a result of sanctions, the oil pipeline from Beira was empty, waiting for the day when Rhodesia became Zimbabwe.

After the break-up of the Central African Federation, certain joint services between the members had for a while continued. These included airways, railways, Kariba and agricultural research. After U.D.I. these, too, broke down, except for Kariba. The airways were most easily sorted out though Zambia got the worse of the division for the workshops were situated in Salisbury. It is, however, simpler to divide planes than rolling stock and lines of railway; moreover, the railway was a huge project. It was run as a unitary system up to November 11th, 1965, and a Higher Authority (appointed by both countries) controlled the Rhodesia Railways Board and met regularly. Rhodesia Railways had borrowed funds on a joint basis (for the two countries) from the World Bank and on the open market.

After U.D.I. the railways were broken up; the workshops were in Bulawayo anyway and most of the rolling stock was seized by Rhodesia. The white mechanics and drivers, for the most part, also opted to work inside white-controlled Rhodesia. The repatriation of these men took some time but all either went or retired. Thus Rhodesia Railways (in Rhodesia) had to grapple with a smaller network, with too many workers, reduced loads and a growing debt; it continued to work because it had the men and material and because the illegal government would not honour the railway's debts. The business of dividing the railway's assets between the two countries has still to be completed; the old company affairs have yet to be settled. The British government undertook responsibility for the repudiated debts but Rhodesia Railways moved further into graver problems in the years after U.D.I. as it could not replace rolling stock, other spares or line as a result of sanctions. Following Smith's 1973 closure of the border with Zambia it was to lose a further large amount of greatly needed revenue. In 1971 the Rhodesian Secretary for Transport reported that the distribution of assets and liabilities of the unitary system had not been resolved: Zambia still refused to come to terms, asking that the assets should be revalued by 30 per cent and be divided equally. In fact 70 per cent of the assets remained in Rhodesia.

Thus the new Zambia Railways had to start virtually from scratch. It had no skilled work force and no workshops except for minor repairs; it had no experienced top personnel and little rolling stock. Management from Sudan was procured but was unsatisfactory: the Sudanese were neither skilled railway people nor trainers. The accident rate soared alarmingly. Zambian staff was dissatisfied; nepotism grew apace and appointments were made that should not have been, resulting in further inefficiency. But the main problem was the appalling loss of rolling stock through ever-increasing accidents. Then Canadians were brought in and they did begin to bring some order into a chaotic situation. Accident rates were extremely high: the Canadians demoted drivers, attempted to deal with drunkenness, which was a major problem, and with union dissatisfaction and the debt burden. No accounts have ever been published. Operational losses were high enough, but the actual losses, taking into account new purchases, were extraordinarily high. Nonetheless, a new workshop was built at Kabwe, and Zambian personnel were constantly on crash training courses.

In the post-U.D.I. situation the Rhodesians had the upper hand as far as transport was concerned and they knew it. They exploited the situation to the full. The copper had to go out from Zambia and seeing the huge obstacles met in attempts to send out copper by road to Dar es Salaam — accidents on the 'Hell Run' were a daily occurrence — Salisbury increased pressure upon Lusaka. Rhodesia demanded currency in advance; she introduced a truck for truck rule which meant that rolling stock going north had to be replaced by southward traffic over the Victoria Falls railway bridge. Rhodesia also refused to service the old loans of pre-U.D.I. Rhodesia Railways. Finally — a portent of things to come — the Rhodesians, afraid that the traffic upon which the railway relied for its prosperity might be absorbed by the alternative routes that Zambia was increasingly anxious to develop, demanded a minimum of 25 per cent traffic of her neighbour's monthly copper output: if this minimum was not met a surcharge was imposed. This latter provision stopped Zambia from carrying out an undertaking it had arrived at in the early days after U.D.I. In 1966 a meeting in Zaire with the Benguela Railway Company had resulted in Zambia agreeing to send a certain amount of copper to Lobito. As a result Tanganyika Concessions, the owners of 90 per cent of the railway which plies between Lobito and the Shaba border, invested in an expansion of capacity as well as a longer programme to modernize the line.

Up to January 1973, however, the Benguela line carried nothing like the expected load though it played an invaluable part in the matter of bulk imports. Until the border closure the Benguela railway carried an average of 15,000 metric tons of copper a month: then in January 1973 this was increased suddenly to 35,000 tons and reached a monthly average of 43,000 tons by the end of that year.

Indeed, the border closure provided a major impetus to Zambia's re-routing plans. Up to that point she was transporting her copper out as follows: 13,000 tons a month through the Benguela railway 800 miles to Lobito; 16,000 tons a month along the 'Hell Run' to Dar es Salaam; some by road to Malawi and thence to Nacala; and the balance through Rhodesia. Following the closure, however, she proceeded to send 10 per cent of her copper by rail and road to Mombasa; 35 per cent by road to Dar es Salaam; 55 per cent by rail to Lobito with the added alternative of some by truck to Salima in Malawi and thence by rail to either Beira or Nacala. These alternatives more than took up the 50 per cent copper freight that had formerly passed through Rhodesia to Beira.

Early in 1974, however, Zambia suffered a further shock with the increase in handling rates at the East African ports. She therefore switched traffic back to Lobito and in June 1974 this route carried virtually the whole of the Zambian production for the month – 53,000 tons. This increase, however, came after the Lisbon coup of April 1974; it should not be assumed that Zambia will ever switch fully or permanently to this route nor in fact that it would be logistically possible to do so. The new political situation that developed in Angola after the Lisbon coup led to the dockers in Lobito striking and otherwise feeling out their industrial strength. A strike in May 1974 was swiftly settled but a work-to-rule exercise was to follow. More labour problems were experienced in 1975 when the political future of the territory was far from clear. Suddenly congestion at Lobito became as commonplace as congestion had been for years at Dar es Salaam. Surcharges by shipping companies sent import costs soaring. Then in mid-1975, as a result of the internecine struggle between the Angolan liberation movements, the railway was closed to Zambian copper, her freight piled up at Lobito and the Zambian government was obliged to declare *force majeure* on its copper consignment deliveries. Zambia has learnt the hard way not to rely upon any one route. At all costs it has to diversify its transport systems. At any one time a particular route may be preferable and the easiest to use for either strategic or logistical reasons – but there must be more than one route.

At the same time it has become obvious that the Zambian domestic network has to be built up to expand local markets. The old line of rail from Victoria to Sakonia (a few miles north of the Copperbelt town of Ndola) has attracted the bulk of the urban population. Most of the important towns are situated along this stretch of rail. It is no accident that the railway passed through Kabwe (formerly Broken

Hill) where the country's main lead and zinc mine is to be found, for the B.S.A. Company's main interest was always in minerals. It is here that the new railway workshop is situated, a huge modern structure covering 20 hectares, completed in 1970, where the country's new diesel engines and passenger and goods trains can be repaired. It includes a training section. Zambia Railways began as a separate entity in 1967 and it extends from Victoria Falls Bridge to the Copperbelt. The towns spanned include Livingstone, Zimba, Kalomo, Choma, Monze, Kafue, Lusaka, Chisamba, Kabwe, Kapiri Mposhi, Bwana Mkubwa, Ndola, Kitwe, Chambishi, Chingola and Chililambombwe. From Chambishi there is a line to Mufulira and from Ndola an access line to Luanshya. Since independence coal deposits have been exploited in the Gwembe valley; as a result a railway spur was built from Choma to Masuku. This gives the country a total of 708 route miles.

It is at Kapiri Mposhi that the great T A N Z A M railway link — the Uhuru railway — terminates its 1,160-mile journey from Dar es Salaam. Developments along the line of rail were already following in Tanzania, where the longer stretch is situated, and in Zambia's under-developed Northern Province following the opening of T A N Z A M in 1975. Kapiri Mposhi had become imbued with new life; so had Mpika, while other areas in the Northern Province as a result of demand and supply will become trigger points of growth, producing food for the railway work force as well as having the new incentive of a nearby carrier to send produce to the market-places of the Copperbelt.

U.D.I., traumatic though it was and immensely expensive for Zambia, also proved to be beneficial to the country. Despite the switching of planned targets, despite the cost of the 'Hell Run', the resulting inflation of imported goods, the unsatisfactory stop-go conditions at Dar es Salaam, which was ill-equipped to cope with a sudden additional traffic load, Zambia moved faster towards self-reliance and autonomy than would otherwise have been the case.

The state company, Industrial Development Corporation Limited, renamed Indeco, took over the functions of co-ordinating strategic problems during the immediate post-U.D.I. period. Firstly, the 'Hell Run' operation became organized. A road transport company — Zambia Tanzania Road Services Limited — was established with Italian partnership and management; and sub-contractors, domestic and foreign, worked with the Zam-Tan owned fleet, on the understanding that they would be phased out when the rail link had been completed. Indeco then set about organizing the import and distribution of petrol. At the same time it entered into a joint venture with the Tanzanian government to build an oil pipeline from the coast to Ndola. Originally the refined product was imported along this route whose first pipeline was finished in a record period of only eighteen months; subsequently, additional loops and other additions were needed to accommodate the increased demand that the refinery had to meet in Zambian growth rates. This refinery, completed in 1973, has a 1.1 million ton capacity and stood Zambia in excellent stead at the time of the 1973 oil crisis. Initially Shell and Agip had equal share supply agreements; later these were replaced by an agreement with Saudi Arabia in 1974.

Apart from oil, import substitution speeded up after U.D.I.: goods previously imported from Rhodesia were either obtained from alternative sources — South Africa, ironically, initially benefitted from this — or where possible products were made in Zambia. Even a copper cable factory was established — Z A M E F A — with great success, becoming an exporter to industrialized countries on a small scale in the early 1970s. Indeco's success during this period laid the foundations for Zambia's programme of 'economic reforms' which intended — and achieved — greater control by the state over the country's economy, maintaining where necessary the foreign partner, but taking for Zambia the majority stake.

On the morning of January 9th, 1973, Ian Smith announced the closure of the Rhodesia-Zambia border and said that by the evening no goods would be allowed to leave or enter Zambia via the Victoria Falls. He followed up this statement within ten hours by saying that copper would be permitted to proceed. Zambia did not make use of this offer; nor of a lifting of the ban three weeks later. President Kaunda announced in March 1973 that Smith was 'too hostile a neighbour' to be trusted again and he said that 'we have decided to take away this trade completely from the southern routes and re-distribute it to other routes'.[3] President Kaunda then set out proposals under which Dar es Salaam was to take most of Zambia's traffic — 43,000 tons a month, followed by Lobito with 35,000 tons of imports; Dar es Salaam would be sent an additional 20,000 tons of copper a month, Lobito 30,000 tons. In the event it did not work out this way. Lobito took most of the load and was able to do so because of its recently increased capacity; work on the modernization of the Benguela line had begun in 1970 and by 1973 the new Cubal Variant (see page 98) could be partly used. By the last quarter of 1974 this engineering work was completed. At the same time President Kaunda outlined a plan to purchase more heavy vehicles to haul loads to and from Dar es Salaam and Mombasa and along the other road route through Malawi to link with Beira and Nacala.

The United Nations came to Zambia's assistance. Reporting to the U.N. Economic and Social Council in July 1974, Sir Robert Jackson (Co-ordinator of United Nations Assistance to Zambia — an Australian) pointed out that direct costs to Zambia of the border closure were estimated over 1973—75 at K34.6 million (£26.6 million); that additional borrowing would cost about K12.40 million (£9.3 million) for the same period; and that recurrent costs which up to June 30th, 1974, had already come to K64.5 million (£48.4 million), would for the whole period under review be an estimated K87.6 million (£65.7 million). In all the blockade would cost Zambia about K186.7 million by the end of 1975. By June 30th, 1974, Zambia had received in cash, soft loans or kind some K40 million (£30 million), to help meet the expenses of the blockade. Sir Robert Jackson appealed for further help.[4]

Even so, the tremendous dislocation in the economy caused by U.D.I. was not repeated in January 1973, difficult though the border closure became. Firstly, Zambians had taken over in the intervening years as administrators in their own right in many if not most strategic roles; secondly, the petrol supply line was secured; thirdly, the 'Hell Run' was then a smooth tarred road; and finally, the United Nations came to Zambia's assistance at once.

Smith's action was foolish. It was attributed directly to the death of two South African policemen in the Zambezi valley and so was designed to ensure South African sympathy. But even in this it misfired. South Africa did not wish for any further limelight to be thrown upon the area where for years attitudes on both sides of the racial and political divide had been hardening. The last thing Pretoria wanted was to arouse fresh global passions and partisanship in an area where, inevitably, she would bear the brunt of such antagonisms. When landmines were exploded on the Zambian side of the border in the months following the closure, general sympathies were with Zambia. When two Canadian girls, tourists in Rhodesia, were shot by Zambian troops in a border incident near Victoria Falls, subsequent Zambian explanations sounded lame and contradictory (she handled the situation badly). But initial indignation gave way to a different realization: that the Zambezi was the front line of a war area and that Rhodesia was the aggressor in it because of the racial policies she pursued.

The United Nations took the view that if Zambia refused to use the Rhodesia Railways route this would help enforce sanctions: the U.N. co-ordinator of aid to Zambia, Sir Robert Jackson, made this point repeatedly. Figures produced by the U.N. showed that Zambia had gone some way towards diversification away from the southern routes, although not enough: 55 per cent of her exports travelled south, over 60 per cent of imports came in through the southern route. By 1973, however, direct imports from Rhodesia had been confined to coke, power and transport; after the border closure only power remained. By the end of the 1970s this, too, will cease when the North Kariba power station has been completed in 1978 to supplement the already existing power supplies from Zambia's Kafue hydro-electric developments.

The immediate question at the time of the border closure was how goods could be re-routed, and later how Rhodesia could be more permanently by-passed. Surprisingly, perhaps, both Rhodesia's allies, Portugal and South Africa, helped in this manoeuvre. There were tremendous obstacles: increased costs; delays; shortages of supplies both for consumer goods and for industry. By 1975, two years later, the breakdowns that many observers had predicted had not occurred. Indeed, indirectly, the border closure may have aided the later re-opening of contact between Zambia and South Africa which first became public knowledge at the end of 1974 with the start of 'détente'.

Rhodesia has existed since U.D.I. by the grace of South Africa (and Portugal) as well as the greed of brokers eager to make quick money from the illegal sale of Rhodesia's boycotted goods. But South Africa was the major prop in the sanctions-breaking operations. Without her aid Rhodesia must have collapsed. After the effective withdrawal of the Portuguese from Africa in 1974, South Africa and Zambia both had to rethink the order and structure of their region of the continent. Zambia did not directly benefit from the border closure as she had done from U.D.I. in the previous decade. The development of the copper industry was slowed down by the blockade, costs of imports increased (aggravated in turn by the world-wide energy crisis). Development, in fact, was retarded. As for South Africa, it had to rethink its foreign policy once its Mozambique front had been turned with the coming to power in that territory of Frelimo. Rhodesia, in these circumstances, was expendable.

Even the new direct rail link between Rhodesia and South Africa, which was frantically speeded up after the Portuguese coup, will not help Rhodesia significantly. Zambia legally is in an uneasy position with regard to its southern neighbour. Under international law land-locked countries must be given the right of transit to the sea. Zambia, however, does not recognize the legality of the Rhodesian regime so that communications between them have been a one-sided affair of public statements — unanswered directly by either side — in which policies have been announced. There was never any direct agreement about the tariffs to be charged by Rhodesia Railways: simply an announcement on a take-it-or-leave-it basis by Rhodesia concerning the minimum tonnage below which a surcharge would be applied. Zambia was not given any options but after January 1973 she decided to leave it.

Zambia's second main alternative for a long time was Angola. During the period after U.D.I. up to 1974 and the Lisbon coup, the colonial policy of Portugal made good relations between Zambia and Angola impossible yet Zambia's increasing dependence upon communications through both Angola and Mozambique also made a total break out of the question. As a consequence Zambian criticism of Portugal was muted, tempered by her need to use the Portuguese-controlled railways.

During 1966 and 1967 U N I T A in Angola was responsible for various attacks upon the Benguela railway. As a result Portugal closed the railway and warned of very serious damage to the economies and life of the Congo and Zambia if support for U N I T A continued. Zambia then expelled U N I T A's president, Savimbi, and his movement was forced to move its office to Cairo. In 1968 in the east of Zambia parts of the great east road bridge across the Luangwa were blown up by Portuguese saboteurs, again emphasizing Zambia's transport vulnerability and her need for Portuguese 'friendship'.

When, in January 1973, Smith closed the border, the Portuguese immediately offered increased rail facilities for Zambia through Angola and so, very rapidly indeed, Zambia was able to increase from 13,000 to 39,000 tons a month the flow of copper out of Lobito. In fact, two years before the Smith border closure, a three-year agreement had been signed in May 1971 between Zambia and Tanganyika Concessions (the virtual owners of the Benguela railway) and this was to be renewed in 1974. Under its terms copper shipped on the railway was not to fall below 20 per cent of monthly output without attracting a surcharge while it would also qualify for decreasing rates if it rose above 22½ per cent — an arrangement that was to work to Zambia's advantage after the border crisis. Then, following the Lisbon coup of 1974, when political changes already foreshadowed a different Angola-Zambia relationship, the Cubal Variant on the Benguela railway was opened: the Variant eliminated a critical bottleneck on the line and so ensured that when the T A N Z A M railway became operational the Benguela would also already be in full operation, meeting many of Zambia's transport needs and doing so more cheaply than the alternatives. It then appeared that the Benguela railway would be a successful competitor for Zambian freight even when T A N Z A M came into full operation.

As far as the route to Dar es Salaam is concerned, the Zam-Tan Road Service is an associated company of the National Transport Corporation which falls under Zambia's Ministry of Transport, Power and Works. The N.T.C. has a third share, Tanzania another third and the Italian management company the remaining third. No transit agreement was made; the same will apply to the T A N Z A M railway, which is a joint Zambian-Tanzanian venture. Nonetheless, rail operations are complex and a Tanzanian Zambian Railway Authority has been established with a co-ordinator appointed by each government to deal with the numerous problems already at issue or likely to crop up, such as the question of land tenure of villages along the route, pay scales, and other practical points of management.

One vital point needs to be stressed: Rhodesia Railways, Zambia Railways, C.F.B. (Benguela railway), K.D.L. (Kinshasa-Dilolo-Lubumbashi) are all part of one system. The railways on an operating level work and co-operate with each other. Consultations are regularly held on such matters as train schedules, diameters of brake-hose, conditions of rolling stock, etc. There is no point in loading or re-loading at border posts for goods traffic. Only passengers suffer from political divisions. No passenger can travel, as he might have done in colonial times, from a ship at Cape Town through to the Transvaal, into Bechuanaland, Southern and then Northern Rhodesia, into the Congo and through to Lobito in Angola — a journey of several weeks. An outdated brochure of Benguela railways praises the line's cuisine and the comforts of seeing Africa from its first-class carriages. The brochure exhorts passengers to embark in a European port, change for example to a railway carriage at Lobito, and travel thence to Elizabethville.

The realities of the 1970s have changed this romantic colonial picture. Passengers on the Benguela line until 1974 travelled only by day and under armed escort — and only as far as Teixeira de Sousa on the border with Zaire. If he so desired the hardy passenger could then cross by foot or car into Zaire to Dilolo and catch a train to Lubumbashi (formerly Elizabethville). From there he would have to change yet again and use some other means of transport since the trains to Ndola had then been reduced to goods carriers only. As for the crossing at Livingstone, this was stopped at the time of U.D.I.: subsequently passengers could walk across the border and possibly board the same train on the other side yet this too was stopped in January 1973.

Given Zambia's many problems, her large copper freight and the consequent revenue to be derived from its transportation, it is not surprising that her neighbours have been willing enough to help her with her transport dilemmas. By 1975 the possibility was growing of new links between Zambia and Malawi. In the period 1964—74 relations between the two countries were hardly easy. There is no rail link between them, and Malawi in transport terms was more isolated than Zambia; she is also far poorer. At U.D.I., however, she offered facilities to Zambia which were then taken up to a limited extent. The Great East Road from Lusaka to Malawi links up with the Malawi railway system and thence to the sea through either Beira or Nacala. From 1970 onwards President Banda became increasingly eager to develop his communications with Zambia. (He had borrowed from South Africa — thus breaking the O.A.U. boycott — for the Nacala link that was opened in 1970.) Thereafter he made plain his interest in Zambia using the line. In a parliamentary speech in 1972 Banda said that Zambia should not place all its eggs into one transport basket — whether this was to the south or the north — but should leave its options open. In this connexion he went on to suggest that Malawi's railway to Nacala should be linked into the Zambian system. During the crisis which followed the border closure, Malawi became an essential part of the contingency planning and re-routing exercise carried out by Zambia. Goods from Beira were sent to Blantyre by train and reloaded on to trucks. Chipata, Zambia's main centre in the Eastern Province, benefited from the new routing just as Livingstone was to suffer a decline in both goods traffic and tourists. Zambia's Namboard (National Agricultural Marketing Board) owns a maize silo at Balaka in Malawi: seed maize, fertilizers and imports are stored there and increasing use was made of this facility after the border closure.

The rail route between Balaka and Salima formed a bottleneck — it was single-track — and Malawi officials have long wanted to upgrade it. Doing so will form part of the scheme announced by Malawi in early 1975 to build a link between the new capital at Lilongwe and Mchinji on the Zambian border in order to link with the Zambian network. Malawi borrowed R19 million (£15 million) from the South African Industrial Development Corporation for these new improvements and extensions.

From the Zambian point of view it made a lot of sense to look at ways of tying in her rail system to that of Malawi. The new Nacala link was to be increasingly used by Zambia after January 1973. But a line 400 miles long is still required to link Lusaka with Mchinji on the Malawi border before the two systems will be connected. Such a connection is needed. The link will greatly ease traffic problems between the two countries as well as encouraging trade. Zambia's Eastern Province is a market for Malawi commodities; some of Zambia's tobacco is processed in Malawi. In fact, Malawi transport officials have long had figures and estimates prepared for the day, which came in January 1975, when the presidents of the two countries should give the go-ahead for such transport co-operation to be put into effect. After years of estrangement, therefore, Malawi and Zambia began to move closer to each other in the mid-1970s. In 1974 President Kaunda paid a surprise visit to President Banda: it was stated at the time that they discussed transport problems. Then in January 1975 President Banda paid a state visit to Zambia: during this, at a banquet in Dr Banda's honour, President Kaunda announced that Zambia would build two new railways. The first would be the long-awaited link to Malawi which, when completed, would bring that country into the network now centring upon Zambia; one effect would be to connect Malawi into the T A N Z A M railway and so ensure it need no longer be exclusively oriented southwards. The other line that Kaunda then announced was to run from Zambia's Copperbelt across its North Western Province and into Angola to link up with the Benguela railway and by-pass Zaire. In November 1975 the World Bank announced that it was to undertake a feasibility study for the Zambia-Malawi railway.

The idea of the second line — a new one to Angola — for which President Kaunda said a survey was to be conducted, is in fact an old one. It dates back, once more, to colonial days when an engineer, Marsland, surveyed the north-western region searching for a way to link Zambia's Copperbelt direct into the Benguela railway without using the route through the Congo. When the line does come into existence it will at the same time provide the communications necessary to open up the mineral deposits that exist in this remote region. At the end of 1975, however, as the civil war in Angola grew more critical, any such schemes had to go into limbo.

There is copper in the region. Indeed, the oldest known mine in Zambia is at Kansanshi, near Solwezi, one hundred miles west of Chingola. It was found during the days of the Scramble for Africa and was then acquired by the Zambesia Exploration Company, an associate of Tanganyika Concessions, the owners of the Benguela railway. The mine presents many technical headaches. It was closed down after flooding in 1957. Modern mining began there in 1908 and continued until 1914; the mine was left dormant until 1927, then started up again only to be deserted once more during the Depression years. In 1974 the Zambian Ministry of Mines announced that Kansanshi would again be rehabilitated at an estimated cost of K45 million (£33.15 million); it had been taken over in 1970 by Nchanga Consolidated Copper Mines in which the State holds 50 per cent interest: N.C.C.M. have issued 51 per cent shares to Mindeco (the State mining corporation) and 49 per cent to Zambia Copper Investment (the Anglo-American Group).

Two other important projects exist in the area. First, Mwinilunga Exploration (1970) Company, a joint Anglo-American/Amax venture, has carried out lengthy prospecting work at Lumwana. The results of their prospecting showed most satisfactory possibilities: a huge low-grade ore body in the area of Chisasa suited to open-pit mining. Estimated reserves are very large and will exceed those at the huge open mine on the Copperbelt at Nchanga although the quality of the ore is not as high. More technical work is required before recovery can begin. The ownership of the mine has also to be worked out: it appears that Mindeco will be the main partner, holding 51 per cent; Roan Selection Trust (R.S.T.) Exploration did the work and it is likely that both Anglo-American and Amax will also be involved. The project received little publicity in the first half of the 1970s while changes were being made in the State participation agreements with the various mining groups; these took from August 1973 to early 1975 to work out.

Second, a Yugoslav company, Geomin, prospected the area and found iron ore. The ore bodies were sufficient to justify the building of a steel works based upon them. This project is to be a joint United

National Independence Party (U.N.I.P.)-Energoprojekt venture and work began on it in 1973. It became the subject of a good deal of controversy both as to site and as to the type of technical process to be employed in the steel manufacture.

This mining background illustrates the purely economic side of a new communications development. The projected line will assist the industrialization of the area; indeed, is essential to it. But it will be more than that, for the link is intended to cross central Angola to join the Benguela railway, probably at Luso. This would then provide a more direct means of transporting Zambian copper from the main copperbelt to Lobito on the Atlantic than the line through Zaire. In geopolitical terms it would mean that Zambia could by-pass Zaire altogether. At least it would give her one more option. The idea of such a by-passing link was discussed by President Kaunda and Lord Colyton of Tanganyika Concessions when the latter visited Zambia with one of his Portuguese directors in 1970. But at that time the political climate for such co-operation was vastly different from that existing five years later: news of the visit in 1970 leaked out and the project was dropped — for the time being. In 1975, however, as the T A N Z A M neared completion, Zambia was nonetheless considering other links that, inevitably, would compete with the Uhuru railway. Although the T A N Z A M has played a notable part in the story of political confrontation in southern Africa that has unfolded since 1965, and its strategic role should continue, its economic one has yet to be proved.

As a result of these re-routing activities — themselves largely the product of the politics of the south — Zambia is becoming the centre of a great railway network. She may, however, face a different sort of problem: that of over-diversification. Zambia's programme of communications' diversification has followed an interesting pattern: she started with too few routes; she had justified political fears of relying upon these alone; so she diversified away from dependence upon communications through the south; having done so she also came to fear the possibility of quarrels elsewhere as, for example, with Zaire so that she then became determined not to commit herself to any one route such as T A N Z A M. This reluctance to depend too much on a single route, no matter how friendly the country through which it passed, led to yet further diversification; the result could be that Zambia finishes up with a highly complicated and under-used network of communications which in pure economic terms is unlikely to pay. Time will show whether all these options can be justified; essentially, however, they were developed in response to political forces.

Notes

1. *First National Development Plan* (Zambia Government Printers) Chapter VIII, p. 39.
2. *Financial Times*, January 10th, 1973.
3. K.D. Kaunda, *A Challenge to the Nation* (Pamphlet), (March 1973, Zambia Information Services).
4. Taken from the text of a statement by Sir Robert Jackson, Co-ordinator of United Nations Assistance to Zambia, on behalf of the Secretary-General, Kurt Waldheim, made before the Economic and Social Council on July 15th, 1974.

Landlocked like Zambia, Malawi has fewer options: she argued after independence that for both geographic and financial reasons she had no choice but to deal with the white south, and she oriented her policies accordingly. Sharing a thousand miles of border with Mozambique — on both her east and west — the key to Malawi's policy after independence (like that of her neighbour Zambia) has been that of transport and communications. She claimed that she could not afford hostility to the Portuguese in Mozambique unless first she had ensured adequate communications through Tanzania. Her main rail link with the outside world was the line from Blantyre to Beira; north the line went to Salima; later it was to be extended to the new capital at Lilongwe, and today it forms the country's main north-south artery.

Shortly after independence Malawi made the decision to go for the Nacala link as her second communications outlet after Beira. There was the pseudo-possibility — never really credible — that a part of northern Mozambique would be added to Nyasaland, so providing direct access to the sea; this expansionist dream might have influenced President Banda's decision to plump for the Nacala outlet. Banda argued that his reasons for a closer association with the white states were two: Portuguese agreement to the Nacala link being constructed and South African readiness to finance it. Somewhat grandiosely, perhaps, Malawi saw herself as the northern focal point of a communications network with shipments from the east and west sections of Mozambique passing through Malawi and on to Rhodesia.

Nyasaland's right of access to the sea through Mozambique was recognized by bilateral treaty in 1890; then by the Beira Convention between Britain and Portugal in 1950. When Banda came to power in the early 1960s, and once he had realized that there would be no early independence for Mozambique, he discussed with Julius Nyerere the possibility of Nyasaland using a southern Tanganyikan port should Portugal close Beira to his country; by 1961 a tentative agreement had been reached for a railway to link Blantyre with the northern shores of Lake Nyasa and then for lake and rail transport to be used on the journey to Mtwara in Tanganyika.

In fact a form of détente with Portugal came about as Nyasaland approached independence and there was talk of Banda visiting Salazar and of the new Mozambique railway from Nacala to Vila Cabral being extended to Malawi. In April 1962, the Portuguese Consul-General in

Salisbury, Dr Pereira Bastos, confirmed that Banda was to visit Portugal; he said 'the harbours and railways of Mozambique . . . are all at Dr Banda's disposal'.[1] It became increasingly clear as independence approached that the country could either orient its communications northwards through Tanzania or build on those already existing southwards through Mozambique.

In 1963 Kaunda and Nyerere were considering a direct Zambia-Tanganyika rail link, which was to be the line of the future T A NZ A M railway. It would by-pass the much longer line envisaged by Banda that had to go from Lusaka to Fort Jameson (Chipata) in Zambia's Eastern Province and then across the border and up north through Malawi and the Kaunda-Nyerere talk of a more direct route between their countries gave Banda the excuse to go all out for the Nacala link across Mozambique to the Indian Ocean.

Nyasaland did, however, send two missions out in 1963: one under Cameron and Msonthi to investigate the Nacala link through Mozambique; and another under Chiume to Dar es Salaam to see about the Tanganyika link. The Portuguese were willing to extend the Nacala line to Nyasaland; Nyerere was prepared to place Mtwara under a joint Nyasa-Northern Rhodesia authority. At first Banda appeared to favour the northern link; in June 1963 Chiume saw Kaunda and they discussed the possibility of a railway from Broken Hill (Kabwe) in Zambia to Fort Manning in Nyasaland and then either round the north of Lake Nyasa or to its northern shores as envisaged in the earlier discussions of 1961. Then the idea was dropped and Banda opted for the Nacala link through Mozambique instead.

It should not be forgotten either how poor Nyasaland was at independence or how much she was then solely dependent upon communications to the south. Back in 1935 at a cost of £3 million the bridge built across the Zambezi into Mozambique had meant that for the first time Nyasaland had a direct rail link to the sea at Beira. To Nyasaland, an imperial backwater until that date, the bridge gave direct access to the sea; in 1936 the railway was extended northwards from Blantyre to Salima on the lake. When independence did come, therefore — whatever future plans might be made — Malawi required the goodwill of Mozambique for access to the sea. Perhaps this was the clinching factor in the decision to develop the Nacala link. Banda said that the Indian Ocean was his country's natural eastern boundary. When the Portuguese controlled Mozambique Malawi had to have good relations with them; after 1974 Malawi was equally constrained — purely for reasons of communications — to have good relations with the freshly triumphant Frelimo. More even than neighbouring Zambia, Malawi requires alternatives. Unlike Zambia, she has no copper and no other major resource that anyone wants; the result is that she herself cannot raise the money for new highways. The exception in her case was that she could obtain loans from South Africa — which she was prepared to do — in the 1960s, long before the 1974 détente exercise got underway. She got the loans for highly political reasons rather than because South Africa saw any advantage to herself in helping Malawi develop alternate communications through Mozambique.

The question of communications, therefore, dominates Malawi's relations with her neighbours. In 1964 a Portuguese mission visited Zomba and in May of that year — six weeks before independence — Banda went on a one-day private visit to Mozambique. When he returned home he indicated that he was going to opt for the Nacala route. He had been told that the northern route could not be economically justified but the decision was as much political as economic. An American transportation survey team had, meanwhile, cautiously approved the Nacala route.

In January 1965 Banda announced as part of his 'Gwelo' development plan that the Nacala link was to be built. By September of that year negotiations between Malawi and Mozambique had begun. By August 1966 Banda could tell his parliament that he had been negotiating with the Portuguese about the Nacala link for three years and that a 'yes' to it had been given in 1965. The line was to run from Mpimbe in Malawi to Nova Freixo in Mozambique, joining the existing line at that point. In 1967 a South African economic team under Dr Rautenbach visited Malawi and endorsed the idea of the Nacala line; they also suggested an extension through to Cabora Bassa. South Africa made Malawi a loan of R11 million for the link, and this was tied to a maximum use of South African material. The contract went to Roberts Construction of South Africa. This deal with the Republic was full of dramatic political consequences for Malawi: it isolated her from black Africa and made hers the first black government to recognize the Republic, while she became the major prize of Vorster's so-called 'outward looking' policy. Banda himself spoke of extending the Nacala-Blantyre railway to Cabora Bassa in 1968; the next year he went further and suggested the system should be extended to join Rhodesia railways at Shamva, so creating a continuous line of rail between Blantyre and Cape Town. More ambitious still, he spoke of his railway plan embracing Zambia and reaching as far as the Congo.

The Nacala link was opened in 1970. The advantages of a further extension through to Cabora Bassa were obvious to the Portuguese up to April 1974 and subsequently were to remain so for the new Frelimo government for Nacala is closer to Cabora Bassa than Beira, and the latter, in any case, is over-used. For Malawi the Nacala route is five times shorter than the earlier proposed northern route through Tanzania and most Malawi exports are currently grown in central and southern Malawi so that Nacala is their best port.

Malawi's second alternative is the road route through the Tete Province of Mozambique to Rhodesia. In 1970 it was agreed that this Salisbury-Blantyre road should be tarred and re-aligned and for this the Portuguese government made available to Malawi a loan of £2.5 million — their first ever aid to a developing country. It was in fact the only Malawi-Rhodesia link and from 1971 onwards it was to be threatened more or less continually by Frelimo forces until the end of the war. As a result transport went along it in armoured convoys and there was a 20 per cent fall off in tourist earnings for Malawi. The road was cut in 1972 to prevent supplies reaching Cabora Bassa; this demonstrated only too clearly to Malawi her vulnerability, her need for good relations with Frelimo and the fact that the Portuguese were losing military control of the area. The fact that the road was cut so soon after it had been tarred also illustrated the political reasons behind the Portuguese loan to Malawi: for Portugal the road was a vital highway in her war with Frelimo and she wanted it to take on comparable importance for Malawi. But by that time the war was already running too strongly in Frelimo's favour and it is unlikely — even had there been no coup in Lisbon — that the Portuguese forces would have retained control of it for much longer.

Up to 1974, therefore, Malawi was wholly dependent upon the goodwill of her white neighbours — the Portuguese in Mozambique — for the transport of her imports and exports. By that year, however, even before the coup in Lisbon she was busy repairing her relations with both Tanzania and Zambia. In 1972 Banda had started the process of mending fences with black Africa: he announced early in that year that the Malawi railway system would be extended from Salima through Lilongwe to the Zambian border at Mchinji. Then in early 1975 Banda went on a state visit to Zambia and Kaunda announced that a new railway line was to be built to link the two countries so that at last Malawi was to be worked into the Zambian system including an outlet, however long, through T A N Z A M railway to the north, so that her total dependence upon southern routes would be broken. Once Malawi is joined by rail into the Zambian system, then both she and Zambia in reverse will have another option each at their disposal. A good deal of Zambian traffic went through Malawi in the post-U.D.I. period: to Salima and then down to Beira or out to Nacala.

Malawi needs the revenue that would result from Zambia using Nacala and she wishes to draw Zambia into her own more southwards-pointing transport network while also obtaining a northern exit for herself. The 1971–80 Malawi Development Plan states: 'Malawi is well placed to become a least-cost link in Zambia's international trade network'. High tonnages will be needed, however, if the link is to pay, and Malawi wants some guaranteed Zambian traffic. Also, for political reasons, Malawi calculated prior to the April 1974 coup in Lisbon that with Zambia using the Nacala link there would be less likelihood of Frelimo cutting the line.

Almost all of Malawi's calculations about communications have been of a highly political nature. At the time of independence in 1964 the rapidly growing black-white confrontation of the area made some form of ideological choice imperative. President Banda decided — and made clear his decision in no uncertain terms — that he was not going to align himself with the black militants to his north; instead he adopted a programme of pragmatic politics of co-existence with his white neighbours. There was a good deal of calculation embodied in the choice. First, Malawi did not have anything comparable to Zambia's copper: she was an infinitely poorer country with far less room for economic manoeuvre. Second, her agricultural potential, especially for example her canned fruits and vegetables, needed a market outlet that Rhodesia and South Africa were more likely to provide than either Zambia or Tanzania and this turned out to be the case. Third, Banda assumed that a static position as a result of confrontation would last for some time and despite dramatic developments in the area he was proved right for the ten years 1964—74. Fourth, large numbers of Malawians worked in both South Africa and Rhodesia, and their remitted earnings were a vital contribution to the economy: in 1969, for example, there were 129,000 men working in South Africa and 200,000 working in Rhodesia.

During the years 1963—65, while he was considering which of the alternative communications routes to go for — north through Tanzania or east through Mozambique — Banda must have examined all these factors and in retrospect it seems clear that his eventual decision to opt for the Nacala route must have been worked out in the following terms: that Malawi was too poor to indulge in confrontation politics; that the white regimes appeared entrenched for the foreseeable future; that they presented the most likely markets for what agricultural goods Malawi had to sell; that large numbers of her adult male population got work in the south anyway; and that by adopting a neutral and in the end friendly attitude towards the white south, Malawi would in fact qualify for financial assistance which she sorely needed. In the event this is what happened. When two days before U.D.I. in 1965 Banda made his famous remark that the Rhodesian army could conquer the whole of East and Central Africa in a week, he was serving notice on black Africa that he did not intend to be part of any O.A.U. confrontation with the south and equally he was appealing to the white-dominated south for friendly relations. He might have infuriated black Africa by his attitude; he also kept his country out of involvements that

would have retarded its development and obtained urgently needed aid
as well. When South Africa made Malawi the loan for the Nacala link it
was not because this would make any difference to her communications;
her interest in the loan was entirely political, for it drew Malawi into the
then developing 'outward looking' policy of Vorster. South African
loans to Malawi for the building of the new capital at Lilongwe and the
Nacala rail link were a small price for Pretoria to pay when in return
she could claim that Malawi had accepted South African policy in
relation to independent black Africa. Banda's famous remark, made in
1966, that he would accept aid 'even from the devil', by which he meant
South Africa, was his political answer to economic needs. In Banda's
estimation the best way of achieving economic development lay in
co-operation to the south; the political price he had to pay for this was
isolation from the mainstream of black African politics during the 1960s
and early 1970s. His policy throughout this period was a classic illustration
of the impossibility of separating economic and political decisions; while
his choice of routes through the white-dominated south, although it
could be presented as an entirely economic decision, was so full of
political significance that it gave to the Nacala link especially an added
political importance that transcended its purely economic value.

By 1975, when the white-dominated status quo of the region was
fast collapsing, Banda was quietly shifting gear with a minimum of
public statements in stark contrast to his almost strident justifications
for looking southwards ten years earlier, mending his fences with his
black neighbours, working hard to come to terms with Frelimo in
Mozambique, and welcoming the proposed rail link with Zambia — that
would at last connect him into the northern network — as though he
had been working for it all along. For successful pragmatic politics this
performance would be hard to beat.

Notes

1. *Nyasaland Times*, April 6th, 1962.

Apart from Dar es Salaam — the terminal point of the T A N Z A M —
Angola and the southern area of Zaire through which the Benguela
railway passes form the most northerly part of the southern Africa
communications jigsaw. Of 43,750 miles of roads in Angola, only
2,500 miles are surfaced. It is the railways that link the major mining
districts of the interior to Angola's eastern neighbours, and most
especially it is the Benguela railway — built between 1903 and 1929
— with its railhead at Lobito that has proved vital for carrying both
Zambian and Zairean copper to the Atlantic for trans-shipment.
International traffic on this railway is a major currency earner for
Angola and the line has proved to be of great strategic importance to
both Zambia and Zaire. Angola also has the Mocamedes railway
which serves the high plateau region of Sa Da Bandeiro.
    Angola has profited especially from the transit of Zambian copper
since U.D.I., which greatly favoured the Benguela railway. The line
runs 838 miles from Lobito to Katanga — now Shaba — and then the
Zambian Copperbelt. Forty per cent of all the traffic is from transit
(that is, not Angolan traffic), while 75 per cent of freight earnings
come from this traffic. Politics and the cheaper rates through Beira
long retarded the growth of the line's use, and it was not until 1965 and
U.D.I. that congestion on the southern routes, combined with political
upheavals, worked in its favour. The railway became even more
significant to Zambia following the January 1973 border closure by
Ian Smith. In January 1975 Kaunda announced that a new line was to
be built from the Copperbelt to Solwezi and west through the new
Lumwana major copper deposits and thence into Angola to connect
with the Benguela railway there, probably at Luso, by-passing Zaire.
Thus, the Benguela railway, whose start was so slow and whose
revenues for so long were inadequate, entered a new period of prosperity and use following U.D.I. to become one of the key economic
and strategic lines of the region.

The history of the Benguela railway
An old public relations handout of the Companhia do Caminho de
Ferro de Benguela (C.F.B., or Benguela Railways Company) states that
'passenger trains with 1st and 2nd class through coaches to Elizabethville
connect at Lobito with the Belgian mail steamers from Antwerp. There

93

is also a through service in the reverse direction . . . ' The period following the independence of the Belgian Congo, however, not only saw the
names of countries as well as towns, stations, and political personalities
change, but witnessed a change in the character of the rail services as
well. No longer was it either feasible or desirable to travel as the
brochure of colonial days alluringly suggests from Europe to the Cape
by boat and rail. The month to six weeks' long journey from Lobito via
Elizabethville (Lubumbashi) through the Northern Rhodesian Copperbelt and down into Southern Rhodesia across the Victoria Falls Bridge,
then through Bechuanaland (Botswana) and into South Africa had
become politically and physically impossible. The one-time unitary rail
system which had emerged after decades of in-fighting in the colonial
period was subsequently broken up again by post-colonial developments
that included the confrontation between black Africa and the white
minority regimes of the south that was to dominate the region in the
1960s and 1970s.

The continuous rail link between the Atlantic at Lobito and the
central and southern African railway systems had been effectively disrupted – at least for the time being. Fighting in the post-colonial Congo
broke the links across the border into Angola. Fears that white
mercenaries would use this route ensured that the break in use continued. By the mid-1970s the rare traveller wishing to cross Angola from
Lobito through Zaire and into Zambia had to alight from the C.F.B.
coach and walk or drive to the border post of Teixeira de Sousa, cross
into Zaire at Dilolo and from there take the onward K.D.L. train to
Lubumbashi where, normally, his journey would end. (The K.D.L. –
Katanga-Dilolo-Lubumbashi line – was previously the B.C.K., which
stook for the operating company, Chemin de Fer du Bas-Congo au
Katanga.) Only on rare occasions is a traveller given facilities to carry
on to Sakania, the border post between Zaire and Zambia where there
are no passenger, immigration or customs facilities.

The link across Victoria Falls bridge was effectively broken by
Smith's border closure in January 1973; even before this, passenger
traffic was undesirable and unlikely as a result of U.D.I. and sanctions.
Rhodesia railways, described in the old brochure as joining up with the
Benguela railway ten miles north of Ndola, has been truncated. The
Zambian section has become Zambia railways, and the Rhodesia rail-
ways end has been increasingly affected by sanctions.

The Benguela railway begins on the Atlantic coast at Lobito and
crosses Angola to the Zaire border on the river Lau near Dilolo, a
distance of 838 miles, and again, as the brochure has it, presented not
only advantages in speed but 'is also of great strategic importance'.
The construction of the Benguela railway made economic sense. The
journey round the Cape to southern and east African ports was cut,

94

and easier access given to ships plying this route in search of traffic from the rich mines of the African interior for which they carried supplies and equipment. Little, in fact, has changed either in strategic or economic targets from the Colonial period. Mineral discoveries were the focal points for the development of communications. Rhodes's Cape to Cairo line was conceived as an imperial dream but carried out in a series of jumps from one mineral find to another: diamonds at Kimberley and gold in the Transvaal pulled the rail line from the Cape into the interior; coal at Wankie in Southern Rhodesia, iron ore deposits at Broken Hill (Kabwe) in Northern Rhodesia (Zambia), copper in Katanga (now Shaba Province of Zaire) as well as the copper of the Zambian Copperbelt diverted the line north.

The idea of cutting out the Cape route from the west was conceived and carried out by Sir Robert Williams. Williams formed Tanganyika Concessions (Tanks) and Zambesia Exploring Company in 1899 to develop a terminus on the southern end of Lake Tanganyika for Rhodes' Cape to Cairo railway and to provide a steamer service on the lake. He was granted a concession by the B.S.A. Company: this allowed Tanks the choice of up to 2,000 square miles of mineral concessions in what is today Zambia. George Grey (brother of Britain's Foreign Secretary, Sir Edward Grey) was sent out to map the concessions area. Sir Robert Williams instructed him by letter: ' . . . on receipt of this letter or as soon after as possible, you will proceed north of the Zambezi, with the view to select 2,000 square miles, and further locate therein 1,000 claims on the terms of the concession granted by the British South African Company . . . '[1] Grey discovered the Kansanshi mine, crossed into Katanga and found the copper deposits there. Following Grey's expedition, Williams obtained a concession from King Leopold to explore mineral deposits in Katanga. In 1906 Union Minière du Haut-Katanga was formed by Tanks and its Belgian associates. Meanwhile Williams also obtained a concession from the Portuguese government for the construction of a railway from Angola to the new Katanga mines. The original concept of the company — Tanks — was dropped after the death of Rhodes. The lake steamer idea also died when a vessel ran aground in a storm on Lake Tanganyika.

After the First World War, Tanks sent expeditions to the Congo-Sudan border area but failed to locate minerals. In the 1930s the company set up a number of companies in Kenya and Uganda, the last of which was closed in 1966. In 1950 Tanks moved domicile from London to Salisbury and in 1964 from Salisbury to Nassau in the Bahamas.

Williams was the first to be made aware of the copper wealth of the Belgian Congo and was a co-founder of the exploiting company, Union Minière du Haut-Katanga. He was convinced that the copper traffic from and the supplies to the mines had to be carried from the west coast. But this was the era of the Scramble for Africa. Rhodes' British South Africa Company (Chartered) was pushing northwards; the Germans were entrenching themselves in South West Africa (Namibia); the Portuguese were staking out Angola for themselves; and King Leopold II was creating — first on paper and then by effective settlement — the Congo. In consequence it was far from a simple matter of raising finance through the City or through European bankers. Williams found himself obstructed by the Colonial Office, by Portuguese opposition to foreign finance, by ambivalent British attitudes to the Germans; by Rhodes' obsession for 'his' railway to the north which he was determined should carry as much traffic as possible to make it pay. Competition for business was complicated by political problems. Smuts in South Africa had his own railway designs. And the copper basin of Katanga, long before the resources across the border in British-dominated Northern Rhodesia were found and exploited, was a hive of conflicting economic, financial and political interests. Thus South Africa, Britain, Germany, Portugal and Belgium all had conflicting interests in the area. In certain respects their interests overlapped; in others they were bitterly opposed.

The greatest difficulties were caused by Leopold's ambitions. The Congo was the king's 'personal' property, which he wished to develop unfettered by other outside interests. Foreigners were to be kept out if at all possible. Williams had acquired his concession in the Katanga area because initially there was great doubts as to the true value of the region. At one time, indeed, Williams had to suffer more than doubts: there were actual suspicions that he was deliberately inflating the value of the copper deposits in order to raise the share value. So that he could exploit the concession area Williams founded his Tanganyika Concessions, still quoted on the London Stock Exchange, whose assets include 90 per cent of C.F.B. Benguela Railways (10 per cent belonging to the Portuguese government). Tanks also still has a 17.6 per cent share in Union Minière: this conglomerate's interests changed fundamentally after Zaire had taken over control of the Katanga copper mines and established its own Gecamines to run them and 'early in 1974 a new agreement was signed between Union Minière and the Government of the Republic of Zaire regarding compensation for the expropriation of Union Minière's former assets in Shaba [Katanga]. At the end of 1974, approximately 60 per cent of the agreed lump sum of Belgian Frs 4,000,000,000 had been paid.'[2] Further, Tanks has a stake in Zambesia Consolidated Finance (Z.C.F.), at one time Zambesia Exploring Company, and then the owner of Kansanshi mines. Today, however, Z.C.F. has very different interests from its original African holdings: it is benefitting from North Sea explorations through Scottish engineering concerns.

Williams suffered so many setbacks in his efforts to finance his railway — which in consequence took many years to build — that he was obliged to dispose of some of his shares in Union Minière. (Now, as then, a 'London Committee' works with the Portuguese Railway Company.) The aim of building a railway in Portuguese territory — with the Portuguese as hostile as the Belgians to foreign interests — could not be reconciled with British interests at the turn of the century. Further, Rhodes' Chartered Company convinced the Colonial Office (though not the Foreign Office) that the line would work against British interests. Rhodesia wanted traffic for Katanga to go through her territory; Smuts wanted South African expansion northwards; Rhodesian farmers wanted to supply the mines of Katanga and Northern Rhodesia; Wankie in Southern Rhodesia wanted to get its coal to the same mining area. In the event, Wankie coal and coke did get there because similar cheap supplies could not be found in Angola or Europe. (After the 1973 border closure, for example, Zambia, which had developed its own — expensive — coalfields in its southern province, had to buy coke from West Germany at three times the Wankie price while Shaba continued to import Wankie coal and coke despite sanctions.) Despite these formidable difficulties, Williams in the end prevailed, partly because he was able to convince the Foreign Office that his line would be constructed with British-made materials; eventually he obtained British government help. The finance saga was lengthy and complex. The period of construction finally began in 1904 and by 1914 over 300 miles of line had been completed. Work was interrupted by the First World War and not resumed until 1920. Then problems of finance again slowed progress so that it was not until August 28th, 1928, that the Katanga border was reached (63 miles further than originally planned because of a colonial land deal between the Belgians and the Portuguese). The final linking of the Benguela railway to the Katanga mines and so the completion of the line was only achieved on March 10th, 1931, when the last tracks were laid 90 miles from Tenke.

At Tenke the line divides. One section travels south through Shaba and via Sakania to Ndola. The other veers north-west via Bukama to the river Kasai. At Kamina the branch line to the north-east begins, linking C.F.L. (Chemin de Fer du Congo Superieur aux Grands Lacs Africains) to Lake Tanganyika.

Benguela Railways is the largest single private employer in Angola with some 13,000 employees. By 1975 its steam locomotives were being slowly replaced by diesels. The company embarked upon a major modernization programme in 1972. In that year a report by consultants (Transportation Systems and Market Research) was accepted by the company. The report covered organization, methods, productivity,

planning and methods of maintenance and renewal of tracks, rolling
stock repair and extension of dieselization; some of the recommendations
were carried out at once. One of these was to construct the Cubal
Variant. The objective of this variant was to provide an 80-mile track of
heavy-duty 90-pound rails, following a new alignment to cut out the
bottleneck where the escarpment suddenly rises 900 metres between
Catumbela and Cubal. The old 60-pound rails could not take the strain
and so, before ascent, trains were divided, only to be joined together
again at the top of the escarpment at Cubal. This was a costly time-
consuming exercise. By the end of 1974, however, the company could
report:

> The opening to traffic of the Cubal Variant on 20th October
> 1974 was unquestionably an important milestone in the life of
> your Company for the investment will significantly reflect on
> the operations of your Railway. Indeed, the Variant shortened
> by 44 kilometres the distance between Lobito and Cubal
> (previously 198 kilometres), it eliminated 70 per cent of the
> curves, increased their minimum radius from 100 to 350
> metres and sharply reduced gradients throughout. The problem
> of line capacity has thus been solved for many years to come
> ... Furthermore, the construction of the Cubal Variant made
> it possible for Diesel traction to be introduced on that particular
> section of the railway ... [3]

The Variant had cost £20 million to construct. The result was a virtual
doubling of the line's capacity, and the fact that part of this route could
be swung into use after the 1973 border closure between Rhodesia and
Zambia, using some of the newly delivered diesel locomotives, helped
Zambia immensely with its traffic difficulties at that crucial time. By
1975, meanwhile, the line to Nova Lisboa had been dieselized; by 1985
it is intended that the entire track to the Zairean border will also have
been dieselized.

Up to the border closure the Benguela railway was carrying about
15,000 tons of copper from Zambia a month and handling 8,000 tons of
her imports. By May 1974 the tonnage had climbed to 45,000 tons of
imports, 35,000 tons of copper and 8,000 tons of lead and zinc exports.
Subsequently the level of traffic dropped due to congestion at Lobito.
For the future the railway will have to take account of T A NZ AM
coming into operation; as the Tanks company report for 1973–74
states: 'The present traffic boom is due to circumstances beyond our
control, and it will not be long before a new competing route is opened
which will link Zambia to the port of Dar es Salaam.'

Nonetheless, Tanks opened an office in Lusaka in the early 1970s and maintained direct contact from it with the Zambian government, as well as having their normal operating links through their agents, Leopold Walford. Directors regularly visit Zambia to confer with officials and ministers and working meetings take place between Zambian railways, K.D.L. and C.F.B. railway officials.

For Angola the importance of the railway cannot be overstated. Apart from providing much-needed jobs and security, the workshops at Benguela and Nova Lisboa offer a base for industrial development. The railway serves the important hinterland south of Lobito which is an agricultural and mining region. The diamond industry is concentrated near Luso and iron-ore mines at Cuima are served by a branch line. The sub-plateau region has sisal plantations and other agricultural products including sugar, all of which need and use the railway.

By November 2001 William's concession is due to the Portuguese government. In 1973 Tanks announced that it had successfully nego-tiated a recoupment of 3,300,000,000 escudos (approximately £60 million) net value of the railway assets, this sum to be paid to share-holders over the remaining concession period. The arrangement was ratified by the post-coup Portuguese government and since then there have been talks between the M.P.L.A. government and 'Tanks'. The M.P.L.A. government takes the view that a private railway in Angola today is an anachronism. The more so in this case since Benguela is foreign owned with a hundred year concession which is not due to expire until 2001. The government has already made this view clear to representatives of Tanganyika Concessions. It seems almost certain that 'Tanks' will surrender the concession to the government sometime between 1976 and 1980.

In January 1975 President Kaunda of Zambia was host to President Banda of Malawi: Kaunda used the occasion of a State dinner to announce two new rail links for Zambia, one to connect the Eastern Province to the Malawi system; the other to connect the Copperbelt directly to the Benguela railway and so on to Lobito.

The latter plan represents one working out of several variants in the Benguela saga. The machinations and intrigues which form the back-ground to the slow construction of the line from Lobito to the border of what is now Zaire are part of the story of the Scramble for Africa and the line has been variously affected during its history by the colonial aims and counter-activities of the British. the Germans, the Portuguese and Leopold of the Belgians. Sometimes financial interests cut across political ones. It must seem puzzling to present-day observers that a British-owned railway company should be operating under a concession in what was Portuguese territory in order to serve as a carrier for traffic to and from Zaire. The railway is of basic value to

Angola, southern Zaire and the Zambian Copperbelt. Because of the many involved interests that surrounded it, the railway took a long time to build, only reaching its end of line at Teixeira de Sousa in 1928 and not including what should long ago have made obvious sense, a direct link to Zambia's North-Western Province. The railway's viability has always depended upon the copper from Katanga (now Shaba) and the Zambian Copperbelt. The tug of interests in this area has been between rail links to the south or to the Indian Ocean ports and this rail link out to the Atlantic.

The Zambia of the 1970s needs the direct link to the Benguela for the precise reason that it was first proposed: to provide a transit route for the country's oldest mine, Kansanshi, in the North-Western Province and to open up the area round this location. Indeed, by 1975, the pressure to do so had become urgent. Following its 1970 mining legislation, the Zambian government acquired the mineral rights throughout its own territory that had been given in colonial days 'in perpetuity' to the British South Africa Company of Rhodes. Once these were acquired the Zambian government offered concessions for explorations, prospecting and mining to interested concerns, including the two major companies which had exploited the Copperbelt mines for decades – the Anglo-American Corporation of South Africa and American Metal Climax. Among the concessions taken up in the North-Western Province was one by the Yugoslav State mining exploration concern, Geomin. Iron ore deposits located around Chisasa, some fifty miles from Solwezi, the provincial capital, aroused the interest of Zambia's President Kaunda who had been looking for a trigger point for development in the region.

The iron-ore deposits will form the nucleus of such a development. Early in 1974, following lengthy negotiations and investigations, it was decided to locate a steelworks and rolling mill near the deposits. A Zambian company, Tika, in which Zambia's national party, U.N.I.P., has the major share, is to operate the project, the machinery and equipment for which is to be supplied by the German company, Demag. Work on the site and mine had been underway since 1972 although Demag only announced its contract in 1974. Major infrastructure, including new transport facilities, will be vital: coal has to be brought to the site from coalfields in the Southern Province, and feedstock (naptha is the basis of the process to be used in the steel production) has to be brought in from the refinery at Ndola while the finished product has to be carried out to its markets.

As for Kansanshi, its history reaches far back into pre-colonial days: it is estimated that it was first worked 1,500 years ago. The mine is again to be exploited. In 1951 Kansanshi Copper Mine Limited began to drain the property, and production began again in 1956 only to stop a year later as a result of major flooding. In 1970 Tanganyika

Concessions sold the mine to Nchanga Consolidated Copper Mines, formerly controlled by Anglo-American and at that time controlled by the Zambian government which had acquired a 51 per cent interest in it as of January 1st, 1970. The technical problems of Kansanshi — situated about one hundred miles west of Chingola on the Copperbelt — have still to be fully solved. But the known ore reserves of 8.48 million tons at a grade of 3.06 per cent copper makes it a necessary part of the Zambian government strategy (as announced in March 1974) to open up the North Western Province, which in the past has suffered from an exodus of its people, especially the younger ones, to the more fertile and more heavily populated regions along the line of rail.

The North Western Province is also of considerable strategic importance to Zambia. This became obvious during the first decade after independence. For example, certain chiefs who had sided with the original nationalist party, the African National Congress, crossed the border into Angola after independence. Lusaka was uneasy about the area for some time because of the proximity of Angola, the blood-ties across the border and a degree of anti-government feeling. Violence erupted in 1971. Officially tribal troubles over the chieftanship were blamed. What became clear to the government was the need to bring the area out of its isolation and backwardness. The new rail line and the development of the mineral resources should do that.

Another security problem in the region resulted from the war in Angola. The province (with its neighbouring Western Province) was the obvious area for transit camps and the passage of freedom fighters into Angola. In the North Western Province for example, some of its troubles were attributed to villagers being terrorized by armed guerrillas. When internal disputes broke out among the leaders of M.P.L.A. (the Popular Movement for the Liberation of Angola) in 1973, some of the fighting took place in the North Western Province, where a camp was taken over by followers of the dissident Chipenda. More complaints of villagers being terrorized by the freedom fighters followed. Since that time Zambia first, uneasily, backed Chipenda and quarrelled with Neto, then reluctantly accepted Neto as the leader of M.P.L.A. again. Yet, when November 11th, 1975, came and Angola became independent, Zambia, like a number of other African countries, did not rush into immediate recognition of any of the groups but watched in frustration as the three liberation movements with their big-power backers fought bitterly for mastery of the rich territory. Only in May 1976 did Zambia recognize the then victorious government of M.P.L.A. Meanwhile Angola's ports and the vital Benguela railway had come to a standstill.

Apart from solving internal problems of control and development, the new link direct into Angola will also lessen Zambia's dependence upon Zaire. Since 1973, when President Mobutu was included for the first time in the regular meetings between presidents Kaunda and Nyerere, there has developed a reasonable relationship between the two countries which earlier had faced a number of points of friction having on the whole poor relations for most of the 1960s. Even so, points of friction remain: these principally concern the pedicle road through that part of Zaire which separates the Copperbelt from the Luapula Province of Zambia, fishing rights, and the definition of borders, as well as the use by Zambia of the Zairean railway system. Zambia Railways has often found that the K.D.L. line inside Zaire formed a bottleneck for Zambian traffic. This is due as much to a shortage of rolling stock as to inefficiency, but lengthy turn-round periods resulted nonetheless. Moreover, the K.D.L. often raises its rates unilaterally without consulting either Benguela or Zambia railways, although regular meetings between the three companies take place.

Whatever relations are enjoyed between countries, it should be a matter of policy — as well as satisfaction — to reduce dependency upon another state for so vital a thing as communications. When the link through to Angola has been completed Zambia will at least find that her transport through to Lobito only depends upon her relations with Angola — and not with Zaire as well. Such a shift has been made possible and politically acceptable as a result of the Lisbon coup of April 1974; once that had taken place it was, in fact, only nine months later that President Kaunda announced the new link would be built. Angola achieved independence on November 11th, 1975, with Portugal proclaiming that it was handing over control to the 'people of Angola', without specifying which organisation, as fighting between the different groups gathered momentum. Yet, already, talks about the link had taken place in the latter part of 1974. These were made easier because of the good relations that had been established between the Zambian government and Tanganyika Concessions earlier in the decade. The likely joining place for the new line and the Benguela railway will be at Luso, but first a new study is to be undertaken to ensure that all the development requirements of the region are properly covered — as far as possible — by the final route that is decided upon for the new line.

During the Angolan war against the Portuguese, Luso was the principal danger spot on the railway and from Luso trains had to be accompanied by armed guards who travelled in turretted armoured cars ahead of the trains. No night travel was permitted between Luso and the border for experience showed that this was the stretch of line most vulnerable to guerrilla attack. This was particularly so because Jonas Savimbi, the leader of the smallest of the freedom groups, U N I T A (United Front for the Total Liberation of Angola) operated from a nearby base. Savimbi had broken away from Holden Roberto's

F.N.L.A. (National Front for the Liberation of Angola) in 1966, and
thereafter operated from within Angola although the F.N.L.A. had
headquarters in Kinshasa and the M.P.L.A. in Brazzaville. Savimbi,
moreover, refused to comply with Zairean and Zambian requests not
to attack the railway. During the decade of struggle the Benguela
railway did in fact come under attack from all three groups at various
times although little material damage was done and few lives were lost.
Because of both Zaire's and Zambia's strategic needs, which depended
in part upon using the Benguela railway, M.P.L.A. and F.N.L.A. did
not often trouble the line. Tanganyika Concessions knew of this
'gentleman's agreement' and was naturally pleased with it and where
possible furthered it with discretion. Dr Neto, the leader of M.P.L.A.,
did make plain that once the T A N Z A M railway had been completed,
giving Zambia an alternative, he would attack the line. M.P.L.A. had
no sense of obligation towards Mobutu's government in Zaire which
had always backed Roberto's F.N.L.A. both before and after the coup
in Lisbon. M P.L.A. had not been allowed either bases or transit rights
in Zaire, a fact that had considerably inhibited its war effort. It was for
these reasons (to obtain the right to use Zaire and recruit among the
half-million Angolan refugees there) that Neto and M.P.L.A. at the
request of the O.A.U. entered into an alliance with F.N.L.A. in
December 1972 — an alliance that in fact was never consummated.
Had the coup in Lisbon not taken place in 1974, so that the war continued, M.P.L.A. would certainly have come to attack the railway. In
a demonstration to show what they might do, M.P.L.A. did launch a
small attack upon the railway on June 14th, 1974, when shots were
fired at a passing train. By then U N I T A's Savimbi had already
indicated his willingness to have talks with the new Portuguese government.

Savimbi, ironically, is the son of a Benguela railway worker — a
station master; shortly after the coup Savimbi wrote to the railway
management in Lobito to say it was then safe for travel on the Luso
stretch of line; he advised the railway authorities to deploy their
troops instead against the F.N.L.A. in the north. The troops in question
were in fact Portuguese regulars although under the command of a
Benguela employee, a retired officer whose sympathies turned out to
be similar to those of the Armed Forces Movement that spearheaded
the Portuguese revolution. Right up to the agreement between the
Portuguese and the three liberation movements of January 1975 the
town of Teixeira de Sousa had the appearance of an army camp: the
border town, still a thriving trading spot, was a centre for the troops
needed to protect the railway line. The troops were also needed to
protect the railway company's plantation workers: the old system had
still not been changed over to diesels so that at various points along

the line eucalyptus plantations needed tending to provide wood to burn. The hacienda workers as well as railway officials at lonely stations were exposed to the possibility of guerrilla attacks and needed security.

The Angola Agreement was signed by the Portuguese, M.P.L.A., F.N.L.A. and UNITA at Alvor in the Algarve on January 15th, 1975; under it Portugal recognized the three movements as the sole legitimate representatives of the people of Angola and recognized the right of the people of Angola to independence. Article Six of the Agreement stated that the Portuguese State and the three Liberation Movements formally affirmed a general ceasefire which was then already being observed *de facto* by their armed forces throughout the territory. Any use of force thereafter other than as decided by the rightful authorities to prevent internal acts of violence or acts of aggression from outside the country were to be considered illicit. Clashes between the M.P.L.A. and the F.N.L.A. developed soon after the signing of the agreement; nor were subsequent reconciliation pacts adhered to and by July 1975 the fighting had intensified, panic spread and refugees began to flee across the country to its ports and airports. The provisional government that had been established in January collapsed in August and it was during that month that Zambia was obliged — because of the increased fighting in Angola — to divert all its copper exports that were then using the Benguela Railway and Lobito away from Angola and declare that it could not meet its delivery obligations.

The activities of the guerrillas in both Angola and Mozambique highlighted the strategic importance of the various railways through those territories. It would have been easy enough in both wars to disrupt the transport systems, and in both wars to do so would have been much to the advantage of the freedom fighters. But the needs of Zambia and Zaire on the west coast, and Zambia and Malawi on the east coast had to be taken into consideration by the Angolan and Mozambican groups. In the case of Frelimo in Mozambique the Rhodesian action in closing the border with Zambia partly solved the question, for once Zambia could not use the line through to Beira Frelimo had few inhibitions about attacking it and this she was just beginning to do seriously in the days before the coup in Lisbon.

The new line planned to run from Zambia's North Western Province into Angola will face stiff competition when it is completed: it will have to compete with the TANZAM railway; it will also have to compete with the Zambian use of the Nacala railway through Malawi when the new line linking that country into the Zambian network has been completed; and once the Rhodesian question has been settled Rhodesia railways (no doubt Zimbabwe railways then) will again be open to Zambian traffic.

In any case the Benguela railway must face a series of problems at Lobito. As a port Lobito handled all the extra loads it had to manage after U.D.I. and again in 1973 the further increases of freight that resulted from Smith's closure of the Rhodesian border with Zambia. But the political change in Portugal immediately affected that country's wage structure and this in turn affected Angola. In May 1974 a railway strike was settled by a wage rise as well as presures exerted by the new administration. The port workers, however, were not as easily pacified. They adopted work-to-rule tactics which throttled operations in Lobito so that by June 1974, for the first time, the port had a line-up of ships waiting for clearance. The syndicates of the Caetano period turned themselves into proper trade unions. One result was that ships stayed outside the limits of the natural harbour and exports of copper from both Zaire and Zambia accumulated to the extent of 40,000 tons which was the equivalent of a month's traffic from Zambia at that time. One shipping line, C E W A L (the North European Conference) imposed congestion surcharges. Following the agreement of January 1975 which led to the participation in the administration of the three liberation movements, there were to be further stoppages. There was hostility to M.P.L.A., and Savimbi's U N I T A was the movement that obtained the confidence of the workers; the dockers returned to work after an appeal from U N I T A.

Any calculations about the use of Angola's railways or ports had to be suspended in November 1975 following the Portuguese withdrawal. Then M.P.L.A.'s main forces were confined to the area of Luanda, the capital, and were preparing to fight off major attacks from both F.N.L.A. and U N I T A: Cuban and Mozambican military detachments had arrived in Luanda to help M.P.L.A. as had Russian technicians accompanying a steady flow of Russian arms. From the south a white mercenary column of mixed South African and Portuguese troops was joining U N I T A in moving against M.P.L.A. From the north the F.N.L.A. forces, backed in their turn by the Chinese, the Americans and Zaire were also moving against M.P.L.A. As a result Africa faced an Angolan crisis on a possibly greater scale than the one that had occurred fifteen years before in the Congo. Yet already from August 1975 onwards, as the fighting became serious, no service was given by the railway to either Zambia or Zaire. Bridges along the line were blown up and in the early stages U N I T A held the greater part of the railway. The fighting over Benguela and Lobito was a see-saw affair and they changed hands several times. By the end of the war, however, officially dating from March 28, 1976, the day the South Africans withdrew from the country, the railway was entirely in the hands of the M.P.L.A. forces from Benguela to Teixeira de Sousa. May 1976 became a crucial month in the history of the railway: the normalizing talks between Mobutu

Neto took place and some Zaire traffic was again being taken on the railway; engineers (mainly Cuban) were fixing damage to bridges and installations; and Zambia had recognised the M.P.L.A. government. Clearly the new government of the M.P.L.A. was as anxious to get the railway working and earning freight charges from its two principal customers as they — Zambia and Zaire — were anxious to use it again.

The collapse of the Angolan route out for Zambian copper simply reinforced the philosophy adopted by President Kaunda, that no matter how apparently friendly a neighbour or efficient at any given time a particular route might prove to be, a landlocked country such as Zambia should develop as many alternatives as possible. Indeed, to prove how the politics of Angola forced a change in the politics of Zambia, ten days after the 'independence' of Angola on November 11th it was reported that for the first time since the border closure of January 1973 Zambian copper was being sent out through Rhodesia again.[4]

Few railways have a more obvious — or, in the context of the events of 1975, more easily illustrated — strategic role than the Benguela through Angola. During the war against the Portuguese the two main liberation movements, M.P.L.A. and F.N.L.A., did not attack it — although it was in their clear military interest to do so — for fear of upsetting their main external supporters, Zambia and Zaire. When Smith tried to force Kaunda's hand over the issue of controlling Zimbabwe guerrillas by closing the Rhodesia-Zambia border in January 1973, the fact that the Portuguese offered maximum use of the Benguela railway to Kaunda — admittedly for their own economic reasons — ensured that Zambia could beat the intended Rhodesian blockade and pursue a more intransigent political line than Salisbury had conceivably imagined. At the start of 1975, as the détente exercise got underway, Kaunda must have felt he was able to work from a position of strength with the newly completed T A NZ A M railway on the one hand and the Benguela railway on the other as the vital means of getting his copper out, so ensuring the safety of his economy. Eight months later in August the closure of the Benguela railway to Zambian traffic fundamentally altered the situation so that by November, whatever difficulties Rhodesia faced (and in the long term they were formidable), Kaunda was no longer working from a position of strength but instead was obliged again to think in terms of using Rhodesia railways. Indeed, the whole history of the Benguela railway, which in economic terms has hardly ever paid its way, has been dominated by strategic considerations and political intrigues.

Notes

1. See Robert Hutchinson and George Martelli, *Robert's People* (Chatto & Windus, London, 1971), p. 109.
2. Tanganyika Concessions Limited, Chairman's Review for the Year, ending December 31st, 1974.
3. *ibid.*
4. *The Times*, November 17th, 1975.

Mozambique's location on the eastern flank of southern Africa gives to her a position of unique importance. Of all Africa's 'outlet' countries, she is the most strategically placed: because of the length of her coast-line, the quality and size of her harbours, and the number of countries whose outlets to the sea she controls.

The long Indian Ocean shoreline of Mozambique is 1,560 miles making the country the natural maritime outlet for Malawi, Zambia, Rhodesia, parts of South Africa and Swaziland. The railways from those countries that pass through Mozambique to its ports are vital — because of the revenue they bring in — to the country's economy. The Lorenco Marques-Transvaal line reached the Rand in 1894 and still serves the area, especially around Johannesburg. There are the Beira to Salisbury and Beira to Blantyre lines; the Maputo to Gwelo line completed in 1954; and the Nacala line to Malawi completed in 1970 — altogether a total of 2,200 miles of track, all of vital economic and strategic value to the landlocked areas of the interior. In addition, major roads begin at both Maputo and Beira, and go to the Transvaal, Rhodesia, Zambia and Malawi. The north of Mozambique as such is badly served by communications; the Zambezi is navigable between its mouth and Tete but has always been under-used. The country possesses three international airports — at Maputo, Beira and Nampula.

Mozambique's harbours, which are of excellent quality in any case, have been greatly enhanced in importance because of the number and extent of the inland territories they serve: for Zambia, Rhodesia, Malawi, Swaziland and Transvaal, Maputo, Beira and Nacala are the nearest ports — and for most of these territories they are the only ports that make economic sense. The country's most important rail lines run from Maputo and Beira. There are three lines starting from Maputo and though together they total a mere 509 miles, they connect Swaziland, South Africa and Rhodesia to the port. These three short lines account for two thirds of all the freight on all Mozambique's railways. The 57-mile Ressano-Garcia line to South Africa is the most significant: minerals go down the line, petroleum up it. The Mozambique Convention of 1909 guaranteed to Lourenco Marques (Maputo) a minimum of 47.5 per cent of overseas traffic for a defined area in the Transvaal in return for South Africa's right to recruit between 65,000 and 80,000 Mozambicans for work in the Rand mines. The port, in fact, has often exceeded this share of traffic because of congestion elsewhere in the South African system.

The guerrilla war in Mozambique led Portugal as late as 1973 to raise loans for a 612-mile highway along the coast from the centre to the north-east running from Vila Paiva de Andrade to the River Lurio so as to link the country's main ports of Maputo and Beira. This was essentially a strategic development dictated by the needs of war.

Maputo is Mozambique's largest port and serves as the principal outlet for Swaziland, the Transvaal and Rhodesia. In 1969 it handled 11 million tons of freight, of which 29 per cent was oil, 31 per cent coal and 40 per cent ore and other goods. It is the best harbour on the whole East African coast and had largely achieved its present form by the time of the First World War. Beira is the sea gateway for Malawi, Zambia (almost entirely prior to U.D.I., still thereafter in part until the border closure with Rhodesia in 1973) and northern Mozambique before the development of Nacala. It also handles some freight for Zaire. The railway from Beira to Rhodesia reached Salisbury in 1899 (narrow gauge) but was later relaid at 3' 6" gauge to conform with the general southern network. It soon thereafter became the principal outlet for Rhodesia. It had long been the main gateway for Nyasaland as well and by 1965, after her independence, was handling 570,000 tons a year for Malawi. The economic growth of Rhodesia produced boom conditions at the port of Beira, although sanctions following U.D.I. led to the closing of the Beira oil pipeline to Umtali. In the days of the Central African Federation Beira was equidistant from London by either the Cape or the Suez Canal routes.

The most recently developed of Mozambique's ports has been that of Nacala. It possesses a large natural harbour while the railway inland from it serves the big northern hinterland of Mozambique and — after 1970 — Malawi. When Malawi links her rail system into that of Zambia Nacala will also be able to serve considerable Zambian needs as well. The Nacala hinterland produces tobacco, cotton, groundnuts and coal at Cuamba. The line from Nacala to Vila Cabral is 496 miles long before the Malawi link is taken into account. The future value of this line is problematical, depending upon economic developments in Malawi in the region of Lilongwe, the new capital; upon the extent to which Zambia uses it — when the new line from Zambia to Malawi has been built; as well as upon political developments in Rhodesia. Even so, in 1972 it was decided that the port's then capacity of 750,000 tons a year was to be doubled by building two new quays and ancillary services. It will become one of the major ports of East Africa. In mid-1973, with increases in rolling stock along the Nacala railway, as well as extensions of facilities in the port, the average turn-round times from Nacala were 23 days whereas for Beira they averaged 42 days. Nacala's importance increased dramatically following the January 1973 border closure by Smith which forced Zambia to look for new routes. This, like so many of the other issues relating to the southern complex of communications, simply highlighted the fact that more often than not it is politics that dictate developments rather than economics.

When Frelimo came to power in Mozambique on June 25th, 1975 (indeed before that date, almost as soon as it was apparent that the new regime in Lisbon was intent on getting out of its African empire), it was already clear that the mere prospect of an independent black government whose leaders happened to be Marxist and totally opposed ideologically to white minority rule was forcing fundamental changes in the policies of the two remaining white regimes in Pretoria and Salisbury. The détente exercise and the beginning of the end for white rule in Rhodesia both got underway following the change of control in Maputo. In this case it was enough that a Frelimo government might close its railways and harbours to either Rhodesian or South African passage — or to both. Salisbury and Pretoria at once had to realign their policies, taking this possibility into account.

For South Africa the change in Mozambique meant the collapse of her policy based upon maintaining minority regimes to her north as a buffer between herself and independent black Africa; this in turn spelt disaster for Rhodesia since her usefulness to the Republic was at an end. It became politic instead for South Africa to embark upon her policy of détente, the principal aspect of which was to force the Smith regime to come to terms with its black nationalists. Smith delayed for as long as he could (he has always been a master of delaying tactics) but Vorster became progressively more insistent that Smith should start to deal realistically with the A.N.C. When as a last gesture of independence Smith incautiously blamed South Africa on British television for delays in the détente exercise he was brusquely summoned to Pretoria and forced to apologize publicly. Not long afterwards (November 1975), Smith let it be known that after scheduled talks with Nkomo of the A.N.C., whatever their outcome, he was going to retire.

Here then is an astonishing story: the mere *possibility* that the new government in Mozambique might close its railways and harbours to goods from South Africa and Rhodesia dictated changes of policy in those two countries. At the same time a totally different twist was given to the politics of the area by the actual closing — as a result of civil war — of the Benguela railway in Angola. Because of that Zambia abandoned her policy of not using Rhodesia railways and again sent her copper out to Beira through Rhodesia giving Salisbury cause for some 'I told you so' elation, although too late to make any real difference to that regime's crumbling fortunes.[1] When news of this Zambian reversal became public in the South African press a veteran Zambian politician, Arthur Wina, a former Minister of Finance, called on the Zambian government to review its policy on the traditional trade route to the south that had been closed since January 1973. He said that as a land-locked country Zambia had to exploit all available outlets to the sea to enable it to manoeuvre economically and reduce the chances of economic blackmail.[2]

Meanwhile, as the civil war in Angola grew in intensity, the South Africans admitted that they had troops involved on the Namibia-Angola border to safeguard their Cunene dam installations and denied (though this was by then public knowledge) that their troops or mercenaries were fighting in Angola.[3] It was almost as though, having decided to come to terms, peacefully and without any attempted intervention in Mozambique, Pretoria then embarked upon a completely contrasting policy. Their limited military intervention in the very different circumstances of Angola a few months later was to put off — for as long as possible — having another hostile Marxist black government on the borders of Namibia on her western flank. Ironically in this fast-moving situation — whatever Frelimo's eventual aims and intentions — Maputo in mid-November 1975 was not only handling vast quantities of minerals being exported from central and southern Africa but Frelimo troops stood guard in the port while South African experts, who were replacing the Portuguese, supervised the shipping operations.[4]

The role of Mozambique in this political-economic puzzle had yet to become clear for the long-term future. It was clarified dramatically with regard to Rhodesia when in March 1976 Mozambique closed her borders with that country. What was apparent were her conflicting economic and ideological interests. Mozambique has still to clarify her attitude to South Africa. As of June 1976, for example, she still permitted the Republic to use her railway outlets.

Against this background in which Mozambique holds the key to much of the political development of southern Africa, the poverty of the country has to be understood. Mozambique is one of the poorest countries on the African continent: it is basically an agricultural country of peasant farmers and it needs all the revenue that it can obtain. A large proportion of the country's income has always been derived from the freight and port handling charges to be obtained from her inland neighbours, especially South Africa and Rhodesia, so that a political decision to close her ports and railways to the regimes of those countries may be an economic disaster for her. On the other hand the new Frelimo government of Mozambique is utterly opposed ideologically to the white policies of Pretoria and Salisbury. Having closed her borders with Rhodesia Mozambique is likely to keep contacts with South Africa to a minimum; even so, the options open to her and the possibilities for change that were opened up when Frelimo came to power have enormous implications for the whole black-white power struggle in

111

the area.

The simplest case concerns Swaziland. Until the Lisbon coup she was to all intents and purposes an economic and political prisoner of South Africa, with three borders surrounded by the territory of the Republic and the fourth sealed in by a Portuguese-controlled Mozambique. With Frelimo in control in Maputo Swaziland could direct all her exports and imports along the short 137-mile railway from Ngwenya to Maputo, and need not have either communications or economic relations with South Africa. Not that this would be easy; but it would be possible. As it happens, Swaziland has one of the most conservative governments in Africa. Nonetheless, the point of an independent Mozambique on her border is that she can, if she so chooses, break away from the constraints that South Africa hitherto has been able to exercise upon her, so that one more of the peripheral countries round the Republic has been freed — at least in a physical sense — of South African domination in terms of communications by the political change that has occurred in Mozambique.

The case of Rhodesia is more complex. Mozambique under the Portuguese was essential to Rhodesia for the breaking of sanctions. Not only did Mozambique provide false certificates of origin and other documents essential to the illegal international marketing of Rhodesia's minerals and other products, but Maputo and Beira were her principal outlets. Even when the U.N.-backed British naval blockade of Beira closed the oil pipeline to Umtali, oil supplies (once organized) were in fact railed up to Rhodesia from Maputo. South Africa's port facilities were already overcrowded so that she could not — and did not wish to — handle more Rhodesia freight. During the build-up of the Frelimo military campaign in 1973 and still more in the early months of 1974, it became apparent that the Beira railway line would shortly be totally put out of action. In anticipation of this — as well as of possible military collapse of the Portuguese throughout Mozambique — Rhodesia rushed through a new railway link from Rutenga to Beit Bridge so as to give her two lines into the Republic (the other being the old 400-mile stretch of Rhodesia railways that passes through Botswana). Rhodesia's great fear both before and immediately after the Portuguese coup was that an independent Mozambique would close all the lines to her and so throw her entirely upon the goodwill of South Africa for her communications. Like Zambia, Rhodesia is loath to be dependent upon any one country for her outlets, however friendly that country may be. She had good reason for such fears as she discovered at the end of 1974 when detente in southern Africa got underway and Salisbury found that she was regarded as expendable by both sides.

By the end of 1974 when South Africa was talking directly with black African states about détente it became clear that in strategic terms the Republic and Mozambique between them held Rhodesia's future in their hands: the Republic because she can exert such powerful pressures upon Salisbury which the latter can partly withstand only if she has at her disposal some alternative outlet; Mozambique because, after the détente exercise had collapsed by the start of 1976, she was in a position to deprive Salisbury of the only viable means of withstanding South African pressures. Frelimo had already committed itself in general terms to applying sanctions to Rhodesia. The Ottawa Commonwealth Prime Ministers' conference in the autumn of 1974 had discussed possible financial compensation to Mozambique if she were to close her railways and ports to Rhodesia to make sanctions more effective. When the new government did assume full power on June 25th, 1975, it had in its power the means if not actually of bringing down the white regime in Salisbury, certainly of crippling it to the extent of making it totally dependent upon South Africa. When détente collapsed and the guerrilla war resumed in Rhodesia early in 1976, Mozambique opened its 800-mile border with Rhodesia to Zimbabwe guerrillas who can now operate from its territory. This move has so extended the guerrilla war that it now makes it almost impossible for Rhodesia's white forces to contain the guerrillas without massive South African support; but already by mid-1975 — long before the border closure — it was clear Rhodesia would not get this support.

Thus, Salisbury is now entirely dependent upon communications through South Africa. She is likely increasingly to find inadequate facilities available to her because of the Republic's own problems of congestion; and she will, therefore, be obliged willy-nilly to obey political injunctions from Pretoria.

From a long term point of view — ideological considerations apart — it is in Mozambique's interest to employ her stranglehold upon Rhodesia to bring about majority rule more quickly since the sooner this takes place the sooner will the economy of Mozambique benefit: not only from a newly booming Rhodesia, no longer internationally boycotted even in part, but also from recapturing a large proportion of the transit trade of Zambia that has been lost over the U.D.I. period. In strategic terms, therefore, the railways that pass from Rhodesia through Mozambique to Beira and Maputo are that country's lifeline — and she does not control them. By closing its border with Rhodesia, Mozambique ensured that the penultimate phase for the Salisbury regime had begun. Already it was clear that one of the diplomatic weapons being used in the détente exercise during 1975 was the likelihood of Mozambique cutting these communications should the Smith regime fail to come to a peaceful accommodation with its nationalists.

Mozambique can also exercise options with regard to South Africa.

Maputo has always been a vital exit for the Transvaal. For political reasons Mozambique may decide to sever all relations with South Africa. However, she badly needs the revenues accruing from South African freight on the railways and use of her port. Mozambique can make an ideological point by denying the port's use to the Republic; she cannot have upon it the kind of effects she undoubtedly will upon Rhodesia, for South Africa has the capacity to develop alternatives — if necessary at great speed — in addition to the fact that at the end of 1975 her new port of Richards Bay on the Indian Ocean came into operation. Hardline ideological supporters of Frelimo excused her post-independence failure to sever all relations immediately with South Africa on the grounds that first she had to make certain of internal control of the whole country — and especially of the south where Frelimo has least influence. In fact Frelimo may well choose to adopt the sort of policy long followed by Seretse Khama in Botswana: that of having no diplomatic relations with Pretoria and publicly opposing its apartheid policies while permitting essential trade. Even a hardline Marxist government in Maputo may well argue that South African use of her rail and port facilities makes a vital economic contribution to her economy — for the time being.

Ideally Mozambique would like, for both economic and political reasons, to see black majority rule in Rhodesia, and she may well use her strategically strong position in relation to Salisbury to help bring this about. Thereafter her communications network and superb harbours could reap maximum advantage from the traffic of Rhodesia, Zambia, Malawi and Swaziland as well as the Transvaal. When, however, as seems likely by perhaps 1980, an isolated and embattled South Africa has withdrawn into a laager, while all the states around her are not only free but hostile, then once more the question of sanctions against the Republic could become a matter of urgency and Mozambique will again be obliged to consider whether or not to restrict South African use of the railway to Maputo — if she has not already done so by then for ideological reasons.

Notes

1. *The Times*, November 17th, 1975.
2. *The Times*, November 18th, 1975.
3. Dr Hilgard Muller speaking in London, November 18th, 1975.
4. *The Times*, November 17th, 1975.

Years after Zambia's independence, when she had embarked upon her policy of disengagement from the south, the southern railway route through Rhodesia still carried 'about one half of Zambia's trade, both exports and imports . . . through Rhodesia's railways and roads . . . ' as President Kaunda was obliged to admit.[1]

As the events of the 1960s revealed the extent to which Zambia was almost totally dependent upon routes to the south, so the strategic and economic implications of the line to Dar es Salaam became ever clearer. The Germans had built a separate network of railways in Tanganyika when it was their East African colony but, as usual, the network was not linked into the countries on the colony's borders and German East Africa as much as any British colony had a separate growth pattern of its own. The main line in the German system ran from Dar es Salaam westwards to Lake Tanganyika; at least from there – across the lake by steamer – the route joined the Belgian line that ran southwards to Katanga.

Zambia's dependence upon the southern communications system after her independence was mainly due to the fact that the territory up to 1924 had been B.S.A. Company property. The company not only ruled in Northern Rhodesia, it also controlled the mineral rights and the railway to the south, and neither the complicated Belgian network in the Congo nor the Benguela railway could compete on satisfactory terms with the line out to Beira. In geophysical and political terms a line from Zambia north-eastwards to Dar es Salaam would break this pattern of growth; this was the basic reason why, when the T A N Z A M railway became a real possibility, the opposition to its construction was to be so sustained.

Rhodes had hoped that his Cape to Cairo railway would pass north-wards through German East Africa along the shores of Lake Tanganyika, and Tanganyika Concessions was created in part at least to build this section of the railway dream. It never did. Plans for a link from Northern Rhodesia into Tanganyika did crop up over the years – long before the independence era – and in 1952, for example, a study ordered by the British Colonial Office was actually completed: it proved the viability of the idea at least from the point of view of terrain but nothing was done. There were too many other routes, too many competitors, too many conflicts.

The worst aspect of colonialism from a continental point of view was the fact that each area developed on its own, without any consideration of comparable developments in neighbouring territories. Southern and Eastern Africa might have been on separate continents for any joint planning that was done. Tanganyika, even after it had come under British administration, was given different treatment from Northern Rhodesia or, for that matter, from Kenya. Resulting inequalities in development showed this plainly enough in the 1960s when the East African Common Market was established. As far as the East African and Southern African railway systems were concerned there was also the physical problem of their varying gauges. Britain fostered self-sufficiency for the different regions with her federation plans — one of the two Rhodesias and Nyasaland, the other of Kenya, Uganda and Tanganyika. The orientation in each case was outwards rather than towards each other, which would have made obvious continental sense. It was simply not considered during colonial times that intra-regional trade could become sufficiently important to warrant the building of the missing link in the railway system — the future T A N Z A M railway; or even, for that matter, better roads between the areas of southern and eastern Africa. Revenue from the carriers of the mineral traffic has always been huge so that it was never in the interests of Salisbury or the government of the Central African Federation to develop any alternatives to the north. In consequence, up to and after her independence, Zambia was trapped, locked in by the Portuguese possessions of Angola and Mozambique to the west and east, by Rhodesia to the south.

It is impossible to assess the importance of T A N Z A M — the Tanzania-Zambia rail route — without taking into account the traumatic years for Zambia of 1965 to 1967. Until Rhodesia declared U.D.I. on November 11th, 1965, Zambia and Rhodesia railways were one system. Then, overnight, the position changed. It was Zambia and not Rhodesia which at first was most adversely affected by sanctions: Rhodesia, for example, was supplied with her oil from South Africa with secret truck convoys bringing it across the Beit Bridge until the operation was exposed by a British journalist, Peter Hazelhurst, in the *Daily Telegraph*. Zambia, which had been a joint user of the Beira-Umtali pipeline and the Umtali refinery, was obliged to rely upon a British-Canadian-American emergency airlift from 1966 to 1967 to keep its economy going. Fuel supplies bounced in leaky drums along 1,100 miles of appalling gravel road — the Hell Run — from Dar es Salaam to Kapiri Mposhi. Because of its distance from markets, Zambia's economy has always been somewhat precarious, often with little margin to spare in the case of communications breakdowns. Her main export, copper, experiences a three-month time lag from refinery to customer; sales range between 50,000 and 60,000 tons a month and the price varies by hundreds of pounds a ton. For example, in the year 1974–75 prices were between a low of around £500 a ton to a high, at one point, of £1,380 a ton. At any one time with three months' output of copper in the export pipeline, many millions of pounds, depending upon the current price, are tied up in cargo. Indeed, in 1972–73 the Zambian Central Bank asked the two mining companies – Nchanga Consolidated Copper Mines and Roan Consolidated Mines – to raise credit on this asset, which they were able to do.

The cost to Zambia of the Hell Run in the early days of U.D.I. – injuries and deaths to people as well as damaged vehicles – was high. After U.D.I. Zambia never returned entirely to the use of the railway system through Rhodesia: one result was an increased time lag for the supply of imports and on certain items, both consumer and raw material, the time from order to delivery could be as much as eighteen months. There were consequent miscalculations concerning import and export flows. Later, there were political complications of a different nature. Following the 1973 border closure, Zambia found that she could no longer import the explosives needed for the mining industry direct from South Africa through Rhodesia – as had been the case until then. Instead, these had to be shipped up to Dar es Salaam and the Tanzanian government – for Zambia's sake – made a special exception and broke the rule of not handling cargo from South Africa. There were many other problems in this period. Often some stocks – for example of flour – dropped so low that the arrival of a wheat ship in Lobito or Beira and the subsequent quick delivery through to Zambia staved off the possibility of rationing perhaps only by a matter of a week or so. The border closure had some costly results: when a shipload of wheat for Zambia was diverted to Dar es Salaam where there are no bulk handling facilities the consignment had to be packed into bags by hand and then reloaded into trucks, adding greatly to the price of what in the end must have been the most expensive wheat Zambia ever purchased. Other similar incidents occurred, all driving home the same point: Zambia's urgent need for a new transport system that would, as far as possible, be under her control and not subject to the political whims of hostile neighbours.

In 1973 the situation was not as serious as in 1966. Nevertheless, essential mining equipment had to be flown in from Johannesburg: special transport aircraft were engaged for the purpose and a regular service was maintained by Hercules aircraft flying between Botswana and Zambia (Alaska Air). Apart from South African imports these aircraft also brought meat from Botswana into Zambia. Politics were always inseparable from the question of communications and after the border closure the Zambians felt that Britain and British companies were at least as anxious as the South Africans for Zambia to accept

117

Smith's offer of using the route south again when he declared the border to be open once more in the February. On Zambia Radio it was said: ' . . . the British want us to begin using the routes through the rebel colony. That is their advice to us. They can jump into the sea. In the past they have tried to lead us into political and economic ruin by advising us against vital projects such as the oil pipeline . . . '[2] Such bitterness applied far more strongly to the Western refusal to assist Zambia and Tanzania in building the TANZAM railway. The two countries certainly tried to get the West to undertake the project first and for Zambia it was not simply a question of either logistics or economics though both were important; primarily, it was a question of strategy and political survival.

On October 26th, 1970, at an inauguration ceremony at Dar es Salaam, President Kaunda said:

'Many things have been said against our railway which', he continued, 'is perhaps one of the most opposed schemes in the world . . . ' He went on to explain that opposition to it came 'from vested interests in white ruled Southern Africa and their supporters elsewhere in the world. Campaigning against the railway had been wide and intense. The railway would be uneconomic, it was argued. We refused to listen. The railway would be too expensive in relation to economic returns, we were advised. We rejected the advice. The railway would take too long to build, it was emphasised rather discouragingly. We ignored the warning. Above all, it was argued that Zambia did not need the railway after all as UDI with all its hardships to Zambia would end in a matter of months.

'Consequently would-be western investors were discouraged — indeed advised — against the formation of a consortium to construct the railway line. But we are fortunate to have friends among them. Indeed, we are not without friends in western Governments who revealed the efforts in some western capitals to sabotage our efforts in bringing this project to reality.

'The old imperial idea that Tanzania and Zambia lay in the British sphere of influence was a determining factor. Governments, prepared to participate in the project, were discouraged because the British Government did not think the project necessary.

'Zambia was to remain dependent on the white-ruled south for the transportation of her exports and imports. We are to be subservient to white domination for as long as it was in the interests of Western Governments regardless of our objectives and interests.

'This is not strange, since historically the then Northern Rhodesia was organised within a framework designed to create a big sphere of influence for Britain and other Western countries. South Africa's ambition is to create a greater Southern Africa under her sphere of influence with Pretoria as the capital. Zambia and other countries north of the Zambezi are still targets.'[3]

Work on T A N Z A M was inaugurated in October 1970. On August 27th, 1973, the tracks crossed into Zambia. In the years before the line crossed into Zambia a good deal of scepticism was forthcoming in the Western press; first that the Chinese could not finance the railway, then that they would not build it properly, finally that they would fall behind schedule. Such suggestions became rarer and more muted as the line progressed increasingly ahead of schedule. The lengthy negotiations with the Chinese had started in 1967: after the agreement had been concluded Chinese ships were to sail regularly into Dar es Salaam harbour bringing workers and equipment. The line was built at a very fast pace initially, until it came to the toughest country on the route in the southern highlands of Tanzania, yet despite some setbacks these were spanned successfully and the 606-mile stretch of the line inside Tanzania was completed. In October 1975 the line was officially declared open although it had been connected to Kapiri Mposhi in Zambia months earlier so that T A N Z A M had reached the old Zambian line of rail to become part of the whole Zambian network.

The objective of T A N Z A M was always political; it will continue to be so after the liberation of Rhodesia and even when major changes have come in the Republic of South Africa itself. The central African plateau needs the additional alternative. In the years to come traffic for the whole area will be directed and re-directed, depending upon the political situation of the day, on the development of access routes and the prevailing situation at the various ports. The priority that Zambia and Tanzania have given to the building of the T A N Z A M railway has often been misinterpreted. In terms of Tanzania's own development her isolated southern region can now be opened up, no matter what external traffic uses the railway, for along its length Ujamaa (self-help) co-operative villages are being established. Further, the 'ghost' stations along the line of 1975, positioned at every eight or so miles will bring new life to the line of rail as they are staffed and come to be regularly used by the people and villages near them — or those who are attracted to the line of rail for the first time. Even during the actual construction period the line had come to be increasingly used. To the surprise of the Chinese work force the Tanzanians swarmed on to trains as soon as these appeared and loaded their baggage, vegetables, chickens and other livestock as well as their relatives, old and young, into available empty waggons. The disciplined Chinese blinked at this instant socialist utilization of their shining new railway — then accepted it.

Many changes have taken place in southern Africa since Kaunda's railway inauguration speech of 1970 yet the basic situation has remained remarkably unchanged. With the T A N Z A M racing towards completion ahead of schedule it was foolish — from a Rhodesian point of view — for Smith to close the border as he did in 1973, so depriving Rhodesia of revenue at least two years before otherwise need have been the case. As it happened, however, there was yet a further irony in the story, for within weeks of the official opening of the T A N Z A M the news broke that Zambia had begun again to use the railway through Rhodesia to get her copper out.[4] One of the basic problems still to be solved after the coming into operation of the T A N Z A M remained the port handling ability of Dar es Salaam. Smith's move in 1973 brought down upon him the anger of both the Portuguese and the South Africans neither of whom had been consulted and both of whom were to suffer financial loss as a result of the Zambian re-routing to the north. Vorster said bluntly that the move had been made without prior consultation; South Africa not only continued to trade with Zambia but actively helped to solve problems of re-routing and by-passing Rhodesia. Here again was an illustration of the convoluted political relationships of the area.

When détente talks began in the latter half of 1974, following the Lisbon coup, African states were under no illusion as to Pretoria's motives: for her it was 'dialogue' under the new name of 'détente' and the Republic's basic motive was the same as ever — to capture as large a part of the trade and economic control of her northern hinterland as possible. The acceptance by Pretoria of the economic fact of the coming of the T A N Z A M simply led her to change emphasis, not her thesis. Developments in all the land-bound countries which use the East African ports will ensure that Dar es Salaam is never the only outlet for Zambia. Her need for other routes will always exist. The building of the T A N Z A M represents for Zambia and Tanzania a victory over both Southern Africa and her Western allies, yet physically Zambia will still have to diversify her traffic as part of her central transport strategy.

Not surprisingly in terms of the history of the T A N Z A M both East African presidents have given much praise to the Chinese. Kaunda has called them true friends in a number of his speeches and said that just as they have respected Zambia's evolving economic and social system so that respect is reciprocated. President Nyerere of Tanzania, speaking at Kapiri Mposhi at what was to be the Zambian end of the line, said in October 1970 that the link had first been considered in 1947 and that in 1952 preliminary reports were completed 'but nothing happened'. He went on to explain that

... we know also that linking Zambia with the north is vital for the security of free Africa — and especially for our two frontier states ... Developments in Southern Africa and particularly the unilateral declaration of independence by the Smith regime in Rhodesia did however make this project very urgent indeed. For a railway link with the port of Dar es Salaam is vital for the full implementation of Zambia's policy of linking herself to the free African states to the north ... despite the fact that Tanganyika was administered by the British Government, no links at all had been created between our two countries; even the road was designed only for light traffic during dry weather. Consequently when the Smith rebellion of November, 1965, was met by a policy of economic sanctions, the most immediate result was grave problems for the newly independent Zambia.[5]

The Tanzanian President acknowledged that 'Tanzania will receive immediate and direct benefit too' and he added that 'the whole of Africa will benefit ... because trade between the different African countries will become easier and thus the development of us all will increase ... '[6]

Concerning fears expressed by the west that the two countries would come under Chinese influence, President Nyerere said that this was a self-revelation of the West. Not only had Britain failed to build the railway between the two countries it had controlled from 1918 to 1961, but the criticism of the Chinese implied that the West Views aid as 'an instrument of domination ... the Chinese have no colonies in Africa or anywhere else in the world; and their present leadership at least is genuinely anti-imperialist; and so are we.' The president concluded that 'a railway is a railway; and that is what we want, and that is what we are being given. But there is something more; this railway will be our railway. It will not be a Chinese railway, because the Chinese are not building a Chinese railway ... '[7]

Throughout these years Zambia was as much afraid of a blocking of its traffic — which did happen in the end (1973) — as of physical attack from the south. But by the time of the border closure her overall situation was fast changing: the Hell Run had been converted into a fully tarred road; on the Benguela railway the Cubal Variant was under construction and the dieselization programme was being carried out, both increasing the line's capacity as Zambia was soon to discover to her advantage; the pipeline for crude oil from Dar es Salaam to Ndola had been in operation since 1968; and above all — in itself a great psychological boost — the T A N Z A M was nearing completion. This was mainly possible because of the close co-ordination of policies be between Tanzania and Zambia. Not only were the two presidents, Nyerere and Kaunda, close friends; their countries' interests complemented each other. Zambia's need for a north-eastern communications outlet was obvious in terms of all the pressures upon her from the south and the complications that attended her western exits.

For Nyerere's Tanzania a closer link for both economic and political reasons with his richer neighbour was important. Although advocating a policy of self-reliance, Nyerere still knew how his doctrine of African socialism could bring hard times to his people. Tanzania, therefore, needs all the co-operation it can obtain from friendly neighbours. The joint venture in the building of the oil pipeline was an important success for both countries. The additional revenue derived from Zambian traffic at the harbour of Dar es Salaam — despite all the problems arising out of the over-extension of the harbour's facilities — was again of great benefit to Tanzania. It was only too obvious that the more permanent arrangements that would follow once the railway had been completed would have an important part to play in boosting the Tanzanian economy. Many further developments hinge on the railway. A typical early example was the building at Ubongo of a special storage depot for Zambian goods, the largest on the East African coast; while Dar es Salaam harbour must continue to expand to deal with the full quota of Zambian traffic that will pass through it as a result of the railway.

A railway into Tanzania's remote southern interior promises quite different economic developments. It will open up the region for agricultural production and mineral exploration and exploitation. There are known iron and coal deposits in the southern Highlands. Tourism, too, may benefit from the coming of the railway.

Not by any means were all considerations economic ones. There existed the need for both countries — with their similar problems and political outlooks — to be able to identify with each other, and this, as much as economic considerations, became a powerful motivation for the line, which developed into a symbol of black solidarity in face of the oppressive white regimes of the south: not for nothing has it been dubbed the 'Uhuru (freedom) Railway'.

Many of these reasons had evolved long before U.D.I. in Rhodesia so that the two presidents announced in 1965 (before U.D.I.) that they had decided to construct the railway to the Tanzanian coast from the heart of Zambia. Offers were invited for assistance in the building of the line. Nyerere put the position bluntly enough when he told a press conference in July 1965 that he was determined that the railway should be built: 'I am prepared to accept money from whoever offers it and see it is built.' The two presidents made some extensive soundings for aid: in the end only the People's Republic of China was prepared to offer support on the required scale, giving what finally amounted to the largest single economic assistance project financed by that country. This Chinese aid, however, only came after a lengthy period of political manoeuvring.

Shortly before Zambian independence a study was carried out by the I.B.R.D. (International Bank for Reconstruction and Development) but its findings were lukewarm: the World Bank felt that there was no need for another rail link in the area which, it suggested, was already sufficiently serviced by rail tracks. This was a curiously naive attitude in view of the complicated politics of the area and the then gathering storm clouds over Rhodesia. The report went as far as to state that even if the then available rail routes should be closed to Zambia there were still sufficient – if costly – alternatives she could use. The report pointed to the problems of trans-shipment which would be caused by the varying gauges in use by East African Railways, the new proposed line from Dar es Salaam to Zambia and the southern African system which included the one then operating in Zambia. The Bank concluded by suggesting that there was little prospect of an expansion of intra-African trade and forecast that a Tanzanian-Zambian line would operate at a loss into the 1990s. It advocated instead improved road facilities.

Subsequently, also in 1964, a United Nations Survey endorsed the views expressed in the World Bank Report and suggested that too much Zambian capital would become tied up in the project which it described as uneconomic. The World Bank report failed to consider the political problems; nor did it take into account Zambia's copper revenues. The strategic need for the railway was entirely ignored.

The two African presidents did not give up simply because of an adverse survey. Then, in 1965, on the occasion of President Nyerere's first visit to Peking, the Chinese indicated their willingness to help with the project. Chou En-lai visited Dar es Salaam in June 1965; a short time afterwards Chinese experts arrived in Tanzania and began a survey of the Tanzanian portion of the proposed line.

President Kaunda was to continue for some time to work on the assumption that western help might be forthcoming. Late in 1965 an approach was made jointly by Zambia and Tanzania to Britain, the U.S.A., France and West Germany, all of whom were invited to examine the scheme. At the same time Britain did, at least, finance a survey by the Maxwell Stamp company. This was still in progress when in November 1965 Rhodesia declared U.D.I. and Zambia's needs suddenly became very pressing indeed. The Stamp survey, when it appeared, was more detailed and positive than the World Bank report; it did consider the political aspects and also the possibility of carrying copper traffic as well as agricultural products that would come from newly developed areas along the line of rail. Its conclusion was that the new line could make economic sense and would in fact complement existing links which, given the swift developments then taking place in the countries of the region, could absorb such an extension of capacity. The report estimated the cost of building the line at $350 million with an additional $33 million for expansion at Dar es Salaam harbour.

Once this report was out Zambia and Tanzania again asked the West for aid. The report was then subjected to criticisms from the World Bank, the African Development Bank and the United Nations Development Programme and the three institutions wanted to undertake yet another study. This idea was rejected by the two governments who then decided to accept the Chinese offer. The volume of Western objections to the railway had their basis in two considerations: that it would detach Zambia from the South African economic sphere of influence; and that it would point — in strategic terms — like an arrow at the white heartland. As Kaunda pointed out, Western objections to financing the line were based all along on 'political and ideological grounds' while in 1969 President Nyerere remarked sarcastically: 'The world has never seen such a profusion of railway projects in southern Africa as those which are now being canvassed — all of them . . . designed to try to stop the Tanzam railway from being built.'[8] It was clear that once the railway had been built it would fundamentally alter the geopolitical relationships of the area.

President Kaunda visited Peking in June 1967 and in September of that year the three countries — Zambia, Tanzania and China — signed an agreement of undertaking in Peking. It was arranged that China would build the railway, finance it, supply technical personnel and train local manpower. A survey followed and the final agreement was signed in July 1970. Thereafter little time was wasted and construction began almost at once. At regular intervals ships laden with materials and men were to arrive at Dar es Salaam where they were given priority in unloading. Before long the Western powers who had turned down the chance — first offered to them — of building the railway and subsequently had treated the Chinese offer as extravagant propaganda now saw the line begin to grow at speed.

A Tanzanian-Zambian Railway Authority was established with headquarters in Dar es Salaam. Work on the line was labour intensive, there was no red tape concerning work permits or import licenses for the Chinese workers or material required for the line. Local labour was recruited and trained on the spot for the unskilled and semi-skilled jobs while others were sent to Peking to study for the more advanced technical jobs. Since the railway started in Tanzania the Tanzanians gained an earlier awareness of it and its possibilities than did the Zambians: it developed into a new and major line of activity as its route reached steadily southwards into the difficult hill country around the towns of Ifakara, Makumbako and Mbeya. For Zambians the first real impact came in mid-1973 when the line crossed the border. Even then for most Zambians (for security reasons that area was restricted) the railway remained remote except for occasional newspaper articles. Only the farmers along its route found immediate cause for rejoicing as demand for their products rapidly increased.

It was hoped in the early days that the five-year construction period estimated in the Chinese plan could be cut to three despite various engineering problems. During 1974, however, it became clear that this would not be the case. Further, even as the line was at last in sight of its goal during 1975, it became increasingly apparent that many problems would remain before it was fully operational. Security has always been a major aspect of the programme, often with amusing side effects, for no figures have ever been released, for example, of the numbers of Chinese working on the line. Thus there has been constant Western — and especially South African — speculation as to how many Chinese (sometimes inflated to a figure of 45,000) there were in Tanzania and Zambia and what they would do when the railway was finished.

Chinese relations with Tanzania have always been good, while Zambia's policy of non-alignment and her forceful stand against racism made her an attractive partner in Africa for Peking. In their turn both presidents Nyerere and Kaunda have shown their admiration for the austere and disciplined Chinese way of life.

The building of the T A N Z A M railway alarmed Rhodesia because of the impending loss of revenue to her railway system and Vorster because of what he saw as an advancing 'yellow peril'; and both of them in more general terms because they saw Zambia slipping out of the white-controlled southern orbit. To the surprise of almost all Western observers, however, the large numbers of Chinese workers on the railway maintained an exceedingly low profile: they did not practice political indoctrination; they did work side by side with African recruited labour. They established neat camps along the route, shied away from contacts with the local people or anyone else while most of them seemed only anxious to return to China and their families. Only on rare occasions was there evidence of a massive Chinese presence: when a ship of workers arrived in Dar es Salaam harbour; or, in 1972, when a troupe of Chinese acrobatic entertainers came to East Africa to perform for the railway workers and their African hosts.

The Chinese ability to work hard impressed all observers. Under the agreement China equipped and financed a single-track line of 1,162½ miles with 147 stations, 300 bridges, 21 tunnels and 2,200 culverts. A ten-track marshalling yards was established near Dar es Salaam and repair yards at Dar es Salaam, Mbeya and Mpika. Railroad sleepers and poles were manufactured at a locally established plant within the maintenance depot at Mangula in Tanzania's Kilombero valley. The finance required to pay for the project came to approximately $412 million, made available to the two African countries through a thirty-year interest-free loan, repayment of which is due to begin in 1983. The two governments agreed to cover local costs through the purchase of Chinese consumer items: there have been problems in this connexion, as insufficient goods of a kind locally needed were available and Zambia fell behind in her annual purchases though she made efforts in 1972 to increase her purchase of such Chinese goods. After the reorganization of Indeco in 1973 Zambia's State corporation did the purchasing from China for its own shops. The most difficult engineering problems for the railway were to be in the region of southern Tanzania, through the highlands; after the Zambian border had been crossed in 1973 the worst technical problems in building the line were over.

In April 1974 goods started moving on the T A NZ A M railway for the first time. Timber was carried to Mwenzo in northern Zambia. Thereafter regular trips were made in order to relieve some of the congestion of traffic at Ubongo. The volume of goods carried, however, was comparatively small and later that year the operation had to be suspended as railway construction materials were automatically given precedence.

In August 1974 during the visit to the railway of a twelve-man Chinese delegation it was announced that construction of the line would be completed in 1975 and that it would become fully operational in 1976. It is expected that ancillary works such as repair yards will all be in operation and mainly handed over to Tanzanians and Zambians by October 1976.

The Lisbon coup of April 1974 not only shifted the balance of power in southern Africa but changed the emphasis of communications there as well. Thus it was no longer an embarrassment for Zambia to use the Benguela railway through Angola as it had been previously although congestion at Lobito became another problem. Then in May 1974 East African Harbours raised the handling and port charges for all cargo, including Zambian freight. There was consternation in Lusaka followed by lengthy negotiations which included a Zambian threat to transfer traffic from Dar es Salaam to other routes. Further problems erupted later in the year. That December Kenya sealed the border with Tanzania: there had been some acrimonious exchanges about road traffic, with Tanzania claiming that heavy Kenyan vehicles carrying Zambian goods were damaging the northern Tanzanian roads. The difficulties were resolved and a compromise solution achieved, but once more Zambia was made aware of her vulnerability as a landlocked state. The incident also highlighted the question of the eventual viability or otherwise of T A NZ A M.

On the one hand regional development and co-operation between Zambia and Tanzania have already taken place as a result of the construction of the line. New skills are being developed and so far at least two hundred Africans from both countries are being trained in China in various aspects of railroad management. New jobs are being created and new areas are being opened up. There will be commercial benefits on both sides of the border. On the other hand the question remains: when majority rule comes in Rhodesia – and still later, when it comes in South Africa – how much will the enormous economic and communications pulls southwards again draw Zambia back in that direction? And how much in consequence could the T A N Z A M – for all the political importance of its building – then become something of an economic liability or white elephant? The T A N Z A M railway was formally declared open on October 25th, 1974, and ironically, less than a month later, the news became public that despite the border closure with Rhodesia, despite everything that had been said, Zambia was again shipping copper through Rhodesia railways and out of the Mozambique ports. No single event could bear more eloquent testimony to the shifting political fortunes of the area than such a volte face by Zambia.

Notes

1. Kenneth Kaunda, *A Challenge to the Nation,*(Zambia Information Services, March 1973).
2. *The Times*, February 8th, 1973.
3. Speech by President Kaunda, October 26th, 1970, issued as *Background 93/70* by Zambia Information Services.
4. *The Times*, 17th November, 1975.
5. Speech by President Nyerere, October 28th, 1970, issued as *Background 96/70* by Zambia Information Services.
6. *ibid.*
7. *ibid.*
8. See Zdenek Cervenka (ed.), *Landlocked Countries of Africa* (The Scandinavian Institute of African Studies, Upsala 1973).

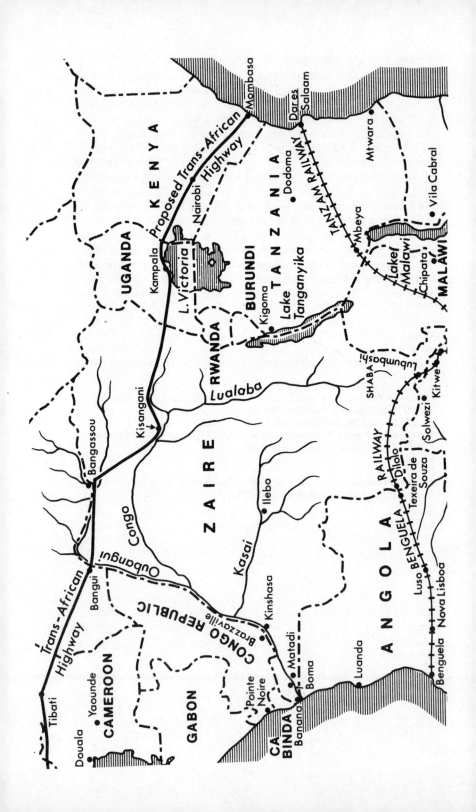

PART TWO   Central and West Africa

12   ZAIRE

Zaire straddles across the centre of Africa: in the south it is linked into the southern Africa communications network; in the north it holds the key to whether or not an effective Trans African Highway whose 1,000-mile central section must pass through Zaire will ever become a working reality. Furthermore, the country itself is continental in size and in consequence faces communications problems on a continental scale.

Minerals, colonial history and geography ensured that Zaire should be drawn into the complicated geopolitical communications network of southern Africa. The wealth of Shaba Province and the connection of the Benguela railway through that region and on to the Zambian Copperbelt make Zaire the northernmost piece of the southern African communications jigsaw.

Only the Sudan in Africa is larger than Zaire. Lying athwart two thirds of the continent in its dead centre Zaire holds a position of enormous strategic importance. Just under a million square miles in area, it has 5,728 miles of borders and only a narrow access to the Atlantic Ocean at the mouth of the Zaire river. To the north Zaire has borders with the Democratic Republic of the Congo, the Central African Republic, and then in the north-east, Sudan; in the east it has borders with Uganda, Rwanda, Burundi and Tanzania; to the south Zambia and then south-west, Angola. Because of its size and the inaccessibility of parts of it, Zaire has to use communications through the territory of her neighbours and they, in turn, do so through Zaire. It is one of the few countries in Africa whose political frontiers, at least in the north and east — with the People's Republic of the Congo, Central African Republic, Sudan, Uganda, Rwanda, Burundi and Tanzania — coincide with natural boundaries. The population in 1975 stood at about 25,000,000, and the country is one of the richest in the world in terms of its mineral resources: both these facts complicate the pattern of its communications. Much of the population is in remote areas, and some of the most valuable minerals — the copper deposits of the Shaba (formerly Katanga) Province — are far inland and away from any port. The development of the Shaba area — in colonial times by the Belgian consortium, Union Minière — and the need to get the copper out to the coast were key factors in the development of its communications network. Zaire depends heavily upon world markets for its revenues.

As with the other great rivers of Africa, the Congo acted as a lure to the nineteenth-century explorers. Stanley's epic journey from

Zanzibar across the continent to the Atlantic in the years 1874—77 made him the first European to explore the greater part of the River Congo (although the Hungarian, Laszlo Magyar, had explored and described the Congo Delta 29 years earlier[1]) and opened up the immensity of the river while also making plain the difficulties of using it as a means of navigation and communication. Later, between 1879 and 1884, Stanley was responsible for the systematic investigation of the Congo region which he helped to open to the predatory colonial grasp of Leopold of Belgium.[2] The mineral possibilities of the Congo Free State (subsequently the Belgian Congo) were early recognized. British pressures from the south meant a long wrangle about borders in the area of Katanga and, indeed, the final settlement of the borders between the Belgian Congo and Northern Rhodesia was not accomplished until 1930: the copper and the other minerals of the Katanga region were to remain in the Congo.

The tangled history of the Belgian Congo, the deliberate Belgian policy of withholding higher education from its subjects and the fierce pursuit by the international corporations of control of the rich mineral deposits of Katanga all laid the groundwork for the tragedies and political disasters of the early years after independence. The result was that the word 'Congo' for the first half of the 1960s came to be synonymous with disaster; it was not until the second half of the 1960s and the emergence of a strong leader in the person of General Mobutu that some form of order and then some form of economic progress could take place. With continuing internal stability in the Congo (renamed Zaire in 1971) it became possible in the early 1970s to re-examine the country's vast mineral wealth and look at the communications needed for its development.

Zaire has 88,473 miles of roads and tracks but only the major roads are surfaced. There are 3,220 miles of railway lines and these are mainly to carry the minerals out from Shaba: these lines link in to both the Benguela railway through Angola to the Atlantic at Lobito; and into the Zambian system at the latter's Copperbelt. Then Zaire has some 7,500 miles of navigable waterways, the principal one being the Zaire itself: this is navigable from Bukama (in the extreme south of the country among the lakes of Shaba) to Kongolo; then, after a break connected by rail, from Kindu to Ubundu; and then when the Stanley Falls have been by-passed, from Kisangani to Kinshasa — the long 1,000-mile main navigable stretch of the river; finally, after yet another haul by rail, from Matadi out to sea. What this means in effect is that for the minerals of Shaba to be transported out the length of Zaire's own river system to the sea beyond Matadi they have to go on a 2,000-mile journey in a great semi-circle through the country — by river and lake to Kongolo; by rail from Kongolo to Kindu; by river again from Kindu to Ubundu;

by rail to Kisangani; by river from Kisangani to Kinshasa; by rail yet again to Matadi; and then finally by ship to sea. The alternative is half the distance — the 1,000-mile rail journey direct from Shaba along the Benguela through Angola to Lobito. Matadi and Boma are Zaire's only ocean-river ports. Of the Zaire's major tributaries, the Ubangi and Kasai are the most used.

Thus, between the mouth of the Zaire and Lubumbashi in the heart of the Shaban mining region five trans-shipments are needed. Altogether there are 2,966 miles of railway line linking the various sections of the river where transport is possible; indeed, it is a main function of the railways to link the navigable stretches of the huge river network. The short Kinshasa-Matadi line is vital, by-passing the early falls and rapids on the river and linking the capital — Kinshasa — to the seagoing port of Matadi. Steamers up to 800 tons can travel as far as Port Francqui (Ilebo) on the Kasai where they connect with another railway line — the old BasCongo-Katanga system — which runs another 800 miles to Lubumbashi in Shaba and thence through to Zambia. Altogether Zaire has five rail systems of which two are connected and two are minor. The 277-mile Matadi-Kinshasa link handles the highest tonnage; there is the short 87-mile line from Tshela to Boma serving the forest and plantation areas; there is the isolated stretch of 425 miles of line in the far north running between Munbere and Bobdo but not connected into the rest of the country's communications; and, finally, the more complex system in the southern part of the country, fanning out north-wards from Lubumbashi: the one line to Ilebo on the Kasai; the second line to Kindu on the Lualaba (part of the overall rail-river transport system); and a branch of that from Kabalo eastwards to Kelemie on Lake Tanganyika. West from Lubumbashi runs the Benguela railway to Lobito and continuing south-east from it the line passes through to the Zambian Copperbelt and so links into the Zambian system. This main Zaire railway system serving Shaba has a total of 1,587 miles of 3' 6" gauge track including all the connexions mentioned above. The Lobito outlet is the most direct for Shaba.

Zaire, therefore, has no satisfactory route out to the coast through its own territory for Shaba's copper and other minerals. This region, in fact, has four possible communications alternatives: first, the long route out through Zaire itself to Kinshasa and Matadi with the various trans-shipments between river and rail which the route involves; second, by rail to Lake Tanganyika, across the lake to Kigoma and then by Tanzanian rail across the whole country to Dar es Salaam; third, south-wards through Zambia and thence to the sea using her outlets (when politics permit) through Beira or South Africa; and fourth, by the Benguela railway to Lobito. When looked at this way, the communica-tions system of Zaire makes the problems of Shaba appear every bit as complicated as if that province were itself a landlocked country.

These complexities make another line in Zaire both economically urgent and strategically vital. This is required so that there can be complete rail transit from Shaba to Kinshasa. A link has been decided upon — the northern of three alternatives — to run just south of the Kasai river from Ilebo (the former Port Francqui) to Kinshasa, some 375–400 miles. This would then join the existing line from Lubumbashi to Ilebo and so make possible a direct rail link from Shaba to the sea. Meanwhile, the Japanese are constructing a railway from the seaport of Banana along the north bank of the Zaire until, near the Inga dam complex, the river is to be bridged — this time by the Americans; then the Japanese-built railway will continue to Matadi and join up with the existing Matadi-Kinshasa line. When completed these two lines will mean that for the first time there will be a continuous rail link from Shaba to the sea entirely through Zaire, so cutting out the need for any river trans-shipment or dependence upon lines through Angola or Zambia. The route would be somewhat longer than that through Lobito but the compensation for Zaire would be the fact that it was entirely within its own territory. A number of people showed interest in building the line from Ilebo to Kinshasa: the Chinese (following Mobutu's growing diplomatic relations with them from 1972 onwards and their growing interest in African involvements); the Japanese; and Lonrho. In the end the contract for the new line went to Lonrho. A railway such as this has considerable political importance for a state as large, diverse and — potentially — divisible as Zaire. Until 1975 and the plans to go ahead with this new line Shaba, with its great wealth in copper, had not relied upon transport through Zaire since the railways through Angola and Zambia have both provided cheaper and quicker ways out for the province's minerals. Clearly this was a factor of great importance when Katanga — heavily backed by outside interests — attempted to secede from the Congo in the early 1960s. Equally clearly it is in Zaire's interest as a unitary State to ensure that strategically she has the internal communications systems to lessen such temptations and possibilities in the future. The new Ilebo-Kinshasa link would serve that purpose.

The river Zaire is the most important river system in tropical Africa. The river itself is the second largest in the world in terms of volume after the Amazon and the entire basin of this vast river lies within Zaire. It has a flow of between 30,000 and 80,000 cubic metres a second and upstream of Kinshasa it is nearly ten miles wide. The huge faults it crosses — first below Kinshasa at the former Stanley Falls (now Ngaliema) and again at points beyond — have meant that despite its vast size the river could not be utilized as a waterway for major communications to the extent that might have originally been thought. The river forms the border between Zaire and the People's Republic of Congo (formerly Congo Brazzaville) and then when the Zaire mainstream moves away from the border this function is played by its principal tributary, the Ubangi — first as between Zaire and the People's Republic of Congo and then between Zaire and the Central African Republic.

Wherever possible water has been used for transport in Zaire and the railways have been built and developed — so far — as accessories to the waterways: to go round the rapids, to link the waterways through areas where they either did not exist or were not navigable. In two stretches of river — the 1,085 miles from Kinshasa to Kisangani and the 378 miles from Kasai to Ilebo — there is great evenness of flow and so these are the most important stretches for transport purposes. The Zaire is fed by streams from both the northern and southern hemispheres so that when one area is dry the river continues to draw water from the other: in consequence, there is always abundant water in the main stretches of the river and so an even flow. The railway system linking the navigable waterways was begun in the last century and the crucial line from Matadi to Kinshasa was completed in 1898. As it is, only 85 miles of the river are navigable from the mouth to Matadi; thereafter there are falls and rapids up to Kinshasa, thus the need for the first railway of 200 miles linking Matadi to the capital and in effect throwing open the whole Zaire area.

The Zaire water system serves a large part of central Africa — not just Zaire itself. First, there is the People's Republic of Congo. Its southern border with Zaire is its only one with a natural boundary: first, the Ubangi river until its confluence with the Zaire, and then the Zaire itself to Songo-Boko, downstream of Brazzaville. Most of the People's Republic of Congo's transport is handled by river. Upstream of Brazzaville a stretch of 800 miles of the Zaire and then the Ubangi links Bangui in the Central African Republic to Brazzaville. This stretch of river is crucial not simply for the economy of the People's Republic but also for its two inland neighbours, the Central African Republic and Chad. Brazzaville itself is a major river port, the trans-shipment point from the river to the railway. The railway — the Congo-Ocean line — completed after thirteen years in 1934, links Brazzaville to the deepwater Atlantic port of Pointe Noire. Although only some 323 miles long, it must be strategically and economically one of the most important in Africa. It is the principal means of export for three countries: the People's Republic of Congo, the Central African Republic and Chad. Pointe Noire has excellent deepwater facilities and handles an annual traffic of about two and a half million tons, nearly two million of this being exports from the three countries. It has been the effectiveness of the river-rail system that has led to delay in the development of any modern road network in the area. If and when the Trans-

African Highway has been fully developed this will provide alternative routes out for the two landlocked countries of Chad and the Central African Republic and so could detract from the importance of this system. This, however, is some way in the future.

The Central African Republic is one of the remotest, poorest and weakest countries in Africa. It covers 243,353 square miles and has a population of 1,640,000 people, an average density of only 6.74 people a square mile, which is far below the continent's average of 4.29 a square mile. Eighty per cent of the total production comes from agriculture and livestock and 90 per cent of all production is for home consumption. As a landlocked country the republic is dependent upon her neighbours for access to the sea so that the river system is of great importance to it. The Ubangi and Sangha tributaries of the Zaire provide access to the sea through Brazzaville and then by the railway to Pointe Noire. Bangui, the capital, is also an important river port handling some 250,000 tons of traffic a year, 75 per cent of this in the form of imports both for the republic itself and for Chad. The 1,763 miles of waterways in the Central African Republic are more important for international than for internal trade and the Republic's central position (just as in the past it made it a crossways for the movement of people) now means it plays a vital part as a communications link for Chad.

Chad is arguably the most remote country in Africa: vast in size (just over half a million square miles), the northern third being desert, the middle section semi-desert and only the south having reasonable rainfall and so agricultural potential. With a total population of 3,800,000 people and a density of 7.576 people to the square mile it is sparsely inhabited and poor. Landlocked, it is dependent for egress to the sea upon communications through the People's Republic of Congo, Cameroon and Nigeria. One of its main routes to the sea, the so-called 'Federal route', is by road from Fort Lamy to Sarh and Bangui (750 miles) and then a further 750 miles by river from Bangui to Brazzaville before using the railway to Pointe Noire. The journey by this route takes from two to three months. Its other routes are out by road from Fort Lamy to Yaounde in Cameroon and then by rail to Douala (there is also a direct, if long, road route to Douala) and both are slow and hazardous. Thirdly, there are routes to Jos and Kano in northern Nigeria by road and thence by rail to Port Harcourt and Lagos. Although the problems of Chad are a good deal more remote from the general river system centred upon the Zaire than are those of the Central African Republic, nonetheless, Chad too is drawn into the system which − slow as it might be − is its most realistic route to the sea for imports and exports.

So large is Zaire that its communications system is equivalent to a good many that elsewhere would be described as international. In addition to serving the best part of one million square miles of territory, the Zaire river system plays a crucial part in the communications of the three countries just examined to its north — the People's Republic of Congo, the Central African Republic and Chad. To the east Lake Tanganyika constitutes an international route of the first order — at least in theory — and may well come to be a great deal more used in practice. The Zaire rail system from Shaba connects with the lake and then across it to Kigoma so linking Zaire into the Tanzanian rail system and out through Dar es Salaam to the Indian Ocean. The eastern regions of Zaire conduct most of their trade through this route which is the shortest to the Indian Ocean and, indeed, the shortest for them to any ocean. Finally, southwards Zaire has growing links through Zambia. Throughout the crisis that followed U.D.I. in Rhodesia in 1965 Zaire continued to import from both South Africa and Rhodesia, and such imports coming through Zambia were a matter of considerable embarrassment to that government during the decade that followed U.D.I. A crucial section of the Benguela railway passes through Shaba in southern Zaire. As a result, that mineral-rich area is linked into the Angolan communications system as well as the Zambian one, giving it access to both the Atlantic at Lobito and, through Rhodesia and Mozambique, to the Indian Ocean at either Beira or Lourenco Marques. Furthermore, it will soon also be able to use the new T A N Z A M route to Dar es Salaam; or the projected Zambian line through to Malawi and thence to Nacala. All these options make the railway complex centred upon Shaba of great strategic importance to the communications network of the area, since it is one of the few important regions in the dead centre of Africa that has direct — if long and difficult — rail communications out in a number of different directions both to the Atlantic and to the Indian Oceans.

The complexities of the Zaire system are a compound of difficult geography, sheer distance, political pulls and the need for more lines. When the projected Ilebo-Kinshasa line is built it will cut a good many knots. And when the Trans African Highway is fully operational — nearly 1,000 miles of it cuts across the northern areas of Zaire — then the country will be in a pivotal position (at least in theory), acting on the one hand as the northern extremity of the southern African complex of communications and, on the other hand, connecting that complex to the new east-west axis from Mombasa to Lagos. The extent to which it might act in this fashion is at least as much a political as an economic question, and the answer must depend upon the policies then being pursued by Mobutu — or his successor.

Mobutu's policies towards three questions — development of the northern areas of Zaire, black-white confrontation in the south, and Angola — give some indications of how this pivotal role may be worked out.

The northern parts of Zaire are remote and under-developed: their communications neglected, their agriculture no longer producing the output of the mid-1960s. It is an area, therefore, that could most benefit from new communications and since of the proposed Trans-African Highway 1,000 miles should pass through the area, it might have been supposed that Zaire would exert strong pressures at least to ensure that that international project went ahead and that she made progress with her own part of the highway. This has not been the case and as at the end of 1975 the Zaire portion of the highway was not only the least developed of the whole but also that part where it appeared little interest was being shown by the responsible government — Zaire (see Chapter 13). Whatever the reasons — internal politics, lack of resources or interest from the top — the fact is that a new Zaire initiative is badly needed if the highway is ever to achieve even its more modest targets in stimulating development.

During the latter half of the 1960s Mobutu showed scant interest in the growing confrontation along the Zambezi, while continuing to trade with both South Africa (despite the O.A.U.-proclaimed boycott) and with Rhodesia, at least some of its sanctions-breaking activities becoming notorious. By the early 1970s, presidents Nyerere and Kaunda were meeting regularly to discuss strategy towards southern Africa. They decided that their policy should be to draw Mobutu into their meetings since they could not afford a country which strategically lay behind them in terms of their confrontation to the south to undermine their efforts: they needed at least the appearance of solidarity. Following his visit to Peking in January 1973 Mobutu was to meet four times that year with Kaunda and Nyerere and at last it appeared that he was joining actively in the question of how to deal with the south. Mobutu, however, has never demonstrated any ideological commitment to solving the race questions of the south and although he was to join in with the other key presidents — Nyerere, Kaunda, Khama and Machel (President-elect) — at the crucial meeting in Lusaka during the October 1974 Zambian tenth anniversary celebrations when the détente exercise was worked out, he later fell out with the other leaders because of his determined support for Holden Roberto's F.N.L.A. in the growing struggle in Angola. And his price for allowing a large expansion of Zambian copper freight to use the Benguela railway after the 1973 border closure with Rhodesia was that Zambia allowed more, not less, goods to come north through Zambia from both South Africa and Rhodesia to Zaire. Mobutu, in fact, has always been entirely pragmatic in his attitudes towards southern Africa; this suggests that he is likely to act as a pressure for the use of whichever route best suits the trading interests of Zaire no matter what the politics involved may be.

Only over the third issue, Angola, did Mobutu show an absolute and continuing commitment to a single policy: to back the F.N.L.A. and Holden Roberto. Rightly or wrongly he has seen an independent Angola under the rule of Roberto as more likely to be friendly — and beholden — to Zaire than the likely alternative through the years of one ruled by Neto's M.P.L.A. By the end of 1975 as Angola was collapsing into civil war this policy was falling in ruins. Then by March 1976 with the triumph of the Cuban-backed M.P.L.A. Mobutu was obliged to come to terms with Neto's government since not to do so would mean the denial of the use of the Benguela railway to Zaire's Shaba province. Mobutu has always shown himself to be a realist. Over Angola he executed a volte face dropping his support for Holden Roberto's admittedly almost totally discredited F.N.L.A. The two sides met in Brazzaville where Mobutu and Neto agreed as follows: that a committee would be established to settle outstanding points between them; that the Angolan refugees in Zaire — perhaps as many as a million — would be invited by the new government to return home but would not be forced to do so; that the 5,000 Katangese mercenaries who fled into Angola a decade ago would be disarmed but not forcibly repatriated to Zaire; that Zaire would withdraw support from the other two movements in Angola — F.N.L.A. and U.N.I.T.A. However Zaire declined formally to recognise the new government in Angola until the Cubans have been withdrawn. Most important for Zaire, arising out of the agreement is the possibility of using the Benguela railway again as soon as it is once more operational.

None of the answers is obvious, but Zaire's pivotal position in terms of communications clearly also depends upon the politics that the country pursues. Under Mobutu they have always been pragmatic.

Notes

1. See Judith Listowel, *The Other Livingstone* (Julian Friedmann, London, 1974).
2. See Richard Hall, *Stanley* (Collins, London, 1974).

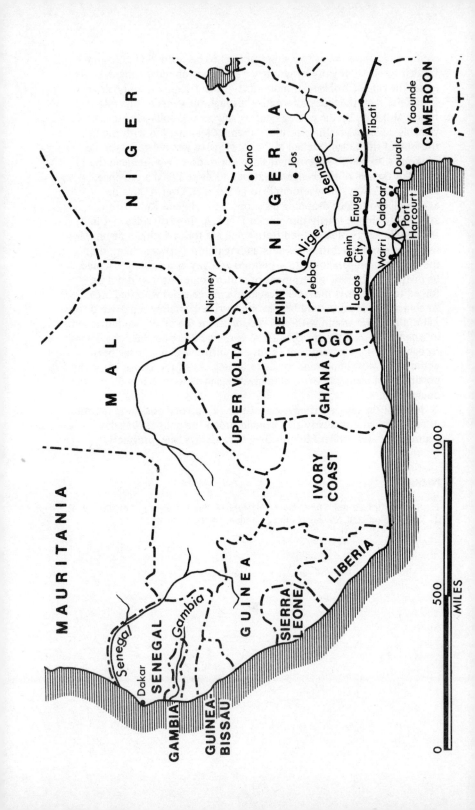

13 WEST AFRICA: E C O W A S AND NIGERIA

The establishment during 1975 of the Economic Community of West African States — E C O W A S — highlighted as never before the lack of any planned development of communications in this vast area of Africa and demonstrated the need for international highways of all kinds: road, rail, air, water. As it is, the coastal states — sometimes alternating Anglophone and Francophone nations — have in common a colonial legacy of vertical communications that run inwards from the coast to secure the interior strategically and economically; hardly ever do these communications run from east to west to connect the different countries in any comprehensive network of highways.

The future of E C O W A S will depend upon the behaviour of its biggest member, Nigeria; and similarly, the development of the whole area must hinge upon the extent to which communications from Nigeria connect it with the other countries in the region. The Trans African Highway starts in Nigeria, and to make economic sense the Trans-Saharan Highway must end there. There is urgent need for a railway to be built along the coast of West Africa to link the main cities and countries: anyone planning the development of the area in a regional sense would have embarked upon such a railway decades ago; colonial development plans, however, did not envisage anything of the sort.

Plans in the abstract may make sense but it is developments in response to urgent needs that produce progress. The realities of development in Nigeria during 1975 meant, for example, unparalleled congestion of her principal ports; so great was the build-up of vessels waiting to be unloaded at Lagos that the Nigerian government initiated serious investigations of alternative routes into the country for its imports. The possibility of whether Dahomey's port of Cotonou could handle Nigeria-bound cargoes was examined — the port is under-used — to see whether these could be sent overland for the final lap of the journey. Similarly, an agreement was reached with Ghana for cargoes consigned to the Nigerian government to be offloaded in Accra and then sent overland. The huge development demands of Nigeria, whose 80 million population must act as the focus for change in the region, have (as a result of oil and the subsequent vast rise in imports) led at last to the kind of international approaches to communications that may begin to make the whole area realize its interdependence for the first time.

Neglect of waterways as a means of opening up the area only started to be remedied during the 1970s. Nigeria's internal waterways have hardly been touched: her great rivers, the Niger and the Benue, have remained a low development priority although in 1972 three of her states — Midwest, East Central and Benue Plateau — did set up a common Water Transport Service. The failure to use the river Gambia as a highway for Senegal has long been a political disgrace. The exploding economic growth of the area in the 1970s highlighted the needs and showed all too clearly the gaps in communications that had to be remedied if West Africa is ever to enjoy any form of cohesive regional development or make E C O W A S more than a pious expression of political hope.

Air communications between the countries of the region are better than those of any other kind: yet all too often these are rudimentary and in some cases non-existent. Telecommunications in the area are primitive. It is easier to make international calls from West Africa to Europe than to connect countries within the region or call long distance within a particular country. Too often the lines are simply out of order for lack of technical services so that the more go-ahead business firms use their own radio-telephone services since they find the State telecommunications systems too unreliable.

Sometimes the economics of desperation produce change. Niger — one of Africa's largest, poorest and most remote countries — must depend upon transport passing through its neighbours to the south. The drought, which affected so disastrously the countries of the Sahel region during the first half of the 1970s, brought out as only a disaster can just how cut off and inadequately provided with any sensible and viable transport system such countries were. Poverty and need helped Niger find new answers: she turned to river transport using the 300-mile stretch of the Niger river that flows through her territory and found that she cut down transport costs by 25 per cent. The country is building up a flotilla of river transport boats following the creation in 1972 of the Niger River and Sea Transport Company (S.N.T.F.M.). Despite the fact that falls have to be negotiated, the river as a means of transport is far more economical than road transport: a ton of freight down the Niger costs 3 Francs (cfa)* a kilometre as compared with 12 to 14 Francs for road freight charges. Until the mid-1970s obstacles to using the Niger river — its falls, rock outcrops dangerous to navigation, and the fact that parts of the river can be navigated only for seven months of the year (otherwise the water is too shallow) — have deterred any serious attempts to employ it as a highway. Then the obstacles were tackled effectively for the first time. S.N.T.F.M. concentrated all its boats and equipment in the lower reaches of the river from the border

* communanté financière africaine = approx. 50 (cfa) Francs to 1 French franc.

with Nigeria through to the sea during the months when the upper reaches are too shallow for navigation.

Three obstacles to effective use of the river had to be overcome. First, the crews of the company's ships had to learn the complications of the river fully. Second, the river service cannot be run up and down the Niger to the sea without negotiating the locks at the Kainji dam in Nigeria: this means that each time a river boat passes through the dam water has to be released at the rate of 1,600 cubic metres a second. The company boats pass the dam every three days: when the water level is high this presents no problem; otherwise it becomes a diplomatic exercise to gain permission for the passage. Third, the bridge at Gaya crossing the river from Niger into Dahomey (close to the Nigerian border) was too low to allow passage to the modern freighters: the solution — expensive as it was — lay in raising the whole bridge. The crews of the company's ships are mixed with the more senior crew members being recruited from Nigeria, although by mid-1975 Niger had some crew members under training in Canada. The history of this small transport company is instructive. During its first year it made only 7 million cfa francs, rising to 38 million in the second year and 78 million (estimate) in 1975; forecasts for its future then suggested a turnover of 200 million cfa francs within a decade. More important, however, has been the determination to overcome obstacles and the fact that in doing so a degree of interdependence between three countries — Niger, Nigeria and Dahomey — was achieved. Numerous comparable exercises will need to be carried out all over West Africa if the region is to realize any true sense of unity.

E C O W A S was created in Lagos, Nigeria, on May 28th, 1975, by a treaty signed by the heads of State or their representatives of fifteen West African countries. It was the culmination of years of talk, manoeuvrings and counter activities and even then it was not certain that all fifteen would subsequently ratify the treaty. The French-speaking members dragged their feet in comparison with the English-speaking ones because it seemed that E C O W A S would render their own Communauté Economique de l'Afrique de l'Ouest (C.E.A.O.) superfluous. Even at the final treaty-signing conference there were five amendments to be dealt with first: but the fact that it did not have an easy birth could mean that its members will work the harder to make E C O W A S succeed.

The treaty itself is not dissimilar in its general provisions from that of Rome establishing the E.E.C.: in theory it will have laid the found-ations for complete integration of the region in little over a decade. The leaders declared that they were conscious of the 'over-riding need to accelerate, foster and encourage the economic and social development of their states in order to improve the living standards of their peoples';

they were convinced that 'the promotion of harmonious economic development of their states calls for effective economic co-operation largely through a determined and concerted policy of self-reliance'; they recognized that 'progress towards sub-regional economic integration requires an assessment of the economic potential and interests of each state'; they accepted 'the need for a fair and equitable distribution of the benefits of co-operation among the States' and finally they affirmed as 'the ultimate objective of their efforts, not only accelerated and sustained economic development of their states but also, the creation of a homogeneous society, leading to the unity of the countries of West Africa, by the elimination of all types of obstacles to the free movement of goods, capital and persons.'

Unexceptional as these sentiments may appear – they are part of the formula attached to all such treaty making – the real question at issue is whether West Africa as a region has greater chances of success than, say, East Africa has so far demonstrated. The treaty continued to lay down that the Community will promote co-operation and development in all fields and especially industry, transport, telecommunications, energy, agriculture, natural resources, commerce, monetary and financial questions as well as cultural and social matters. It has comparable institutions to the E.E.C.: the Authority of the Community (made up of heads of state); a Council of Ministers; a Tribunal and Technical and Specialized Commissions. The decisions of the Authority of Heads of State and Government are 'binding on all institutions of the Community'. The Authority has to meet once a year; chairmanship is by rotation. The Executive Secretariat – the working centre of the Community – is headed by an Executive Secretary appointed to a four-year term which is renewable. The treaty set fifteen years (i.e. 1990) as the period in which total liberalization of trade and the establishment of a Customs Union is reached. Special attention is given under article 14 to the checking of smuggling, something endemic to the region, and – in a sense – one of the most encouraging indications that the community might work: the size and extent of smuggling operations across the various borders indicates that the people of the region have long had a pattern (as well as a determination to continue it) of trading with each other.

Like many such treaties, that setting up E C O W A S expresses a wish, and some members of the community are clearly more enthusiastic than others. The great danger to its success is likely to be the preponderant position of Nigeria: she has 80 million of the community's total population of 120 million and in power and economic terms is the area's giant. If she can tread sufficiently gently so as not to undermine her smaller partners the experiment might work; if she does not, then the others are liable to draw back in the fear that otherwise their economies will be swamped or at least pulled too lopsidedly towards the magnet of Nigeria.

There are, however, sound historical reasons why E C O W A S might thrive. It can be argued[1] with considerable force that the community is the re-creation of something that flourished in the past. In the first place appeals to the past have powerful influence in West Africa. Secondly, the countries concerned — the fifteen of the 'bulge' — have made a market in the past and should be able to do so again. Through many centuries there has been long-distance trade outwards from this area. The trans-Saharan caravan route to West Africa is of great antiquity: it reached a flourishing peak in the fifteenth and sixteenth centuries, although it had dwindled to almost nothing by the colonial era. Islam, especially, tends to be a unifying force in the area and the barriers erected by colonialism have long been ignored by the people living on the various borders, who freely pass back and forth and trade as they will — hence the large-scale smuggling. Some of the ancient West African empires such as Ghana, Gao and Kanem straddled some of the modern frontiers of the community. Hausa is spoken in half the states of the region and other languages certainly cross more than one modern frontier. And so, despite the boundaries left behind by imperialism, there are a number of older historical factors making for regional unity. The treaty could help bring these out to the mutual advantage of the whole area. To do so effectively, however, it will require some rapid development of better communications.[2]

There was a good deal of optimism expressed when E C O W A S was launched but expressions of political intent should not blind the leaders of the fifteen nations to the size of the problems they have to overcome. The area concerned covers nearly 2.5 million square miles and embraces fifteen countries of widely varying cultural backgrounds which in turn are further complicated by the fact that half are 'Francophone' and the others 'Anglophone' territories, while the largest of them all — Nigeria — dwarfs the development possibilities of the rest. No matter how softly Nigeria treads she is bound to give rise to fears that she will dominate the community for her own ends at the expense of her weaker partners. Aware of these difficulties the members, nonetheless, followed up the signing of the treaty in the middle of 1975 by various ceremonies and other expressions of determination to make it work. E C O W A S will have a great deal to overcome between Anglophone and Francophone Africa: the former countries are all members of the Commonwealth; the latter gained early access to the E.E.C. through the Yaounde agreements; and both groups had astonishingly few contacts with one another throughout the colonial era and — unsurprisingly in view of the way communications had been developed — few enough in the years that followed.

In view of these problems it was encouraging that following the May agreement West African Ministers of Transport held a meeting in Abidjan at the request of the O.A.U. and the E.C.A. to work out plans for the integration of the West African transport system: such an effort is greatly needed. The trouble with current communications patterns for the whole region is that they were designed to ensure contact with the outside world rather than across borders with the other countries of the region: this fact alone ensures that E C O W A S has major problems to overcome. The conference took decisions concerning five main categories of communications: road transport, railways, maritime transport and coastal shipping, inland waterways and air transport.

With regard to roads the conference decided that within the sub-region frontier formalities should be standardized to allow free circulation of vehicles and goods across borders on a reciprocal basis. Uniform standards are to be adopted for roads and bridges in the region. On the question of railways the conference recommended that the members should work for the establishment of inter-African linkages by introducing uniform standards when new lines are constructed or when rolling stock is replaced; by developing links between the present unconnected lines where this can be justified; and by working especially towards opening up the landlocked countries of the region.

On coastal shipping and maritime transport the conference made plans for the purchase of ships for the development of merchant fleets and decided to set up an African Maritime Conference for West and Central Africa. It also undertook to study the establishment of multinational shipping companies for the region and decided upon the creation of group freight bodies as well as national maritime insurance. Most of these shipping moves are in their infancy: at least the initial steps were taken towards a group rather than an individual country basis.

The basic question concerning inland waterways was how technical improvements of the physical infrastructure of rivers and lakes could be implemented.

Finally, far-reaching decisions were taken with regard to air transport towards the establishment of a Pan-African Air Freight Company. There is great need for rationalization of all the interlocking services of the region — or in many cases — action to make them interlock. The goal of eventual integration of the air services was expressed though how long before it comes into effect remained very much an open question.

So far almost all action has been theoretical. The E C O W A S treaty followed by this conference at least did two things: recognized the vital need for a common system for the area, and established the necessary committees to start the detailed work that will eventually produce common action.

So important is Nigeria to the whole region and so typical of its problems are her communications shortcomings that West Africa's problems can be aptly illustrated in terms of the one country. Hardly anything in Lagos worked properly in the mid-1970s: the port was unbearably congested; the roads often presented a solid and permanent traffic jam; telecommunications were abysmal. Indeed, as oil brought wealth and accelerated development to Africa's most populous country this simply served to demonstrate just how bad or non-existent proper communications of almost all kinds were within the country itself let alone with its near neighbours. Nigeria's infrastructure remained colonial, geared to slow-moving and small-scale developments long after independence had forced the pace of progress. Then oil produced a boom that outstripped anything the infrastructure could handle. Nigeria's roads have received by far the most attention from her government; her railways are so outdated and rickety that they require almost total rebuilding; her internal air services — let alone the external ones — are notorious for management problems, lack of modern air-craft and poor ground equipment; her ports constitute one of the great economic bottlenecks of the world. The 1975–80 Third Plan set aside vast sums to be used on modernizing the communications infrastructure — and not before time: N4,000 million (one naira = 70p) was allocated for transport, mainly roads; N400 million goes to modernizing the railway; air transportation was allocated N390 million; while the Nigerian National Shipping Line is to acquire nineteen new vessels to enable it to carry 30 per cent of the nation's sea-borne traffic. N774 million was set aside for the improvement of telecommunications.

So bad had port congestion become by mid-1975 that a Nigerian paper could argue plausibly that for all practical purposes the country had become landlocked: it said this at a time when more than 240 ships were waiting in the lines to unload at Lagos and suggested that Nigeria look in earnest at the development of the Trans-Sahara route so as to link its major cities with Algeria's ports.[3] It also suggested that Nigeria had greater reason to participate in the Trans-Saharan route than in the Trans-African Highway. Symptomatic of internal transport problems was the fact that while by 1975 Nigeria had moved into seventh place in the world as an oil producer, so bad was her distribution system that periodically during that year there occurred fuel shortages up country — because supplies did not get through regularly.

In terms of the country's roads — and it should be remembered that Nigeria is the size of the British Isles, France and the Benelux countries combined — there are 37,500 miles of which a quarter are tarred. The main trunk roads are of high quality; the lesser and rural roads need much development. In most cases only canoes for transport are used on the huge rivers. Despite a natural 'continental' need for internal air

145

transport, it remains easier to fly from Lagos to London than from Lagos to places within the country.

For years Lagos, the gateway to the largest market in Africa, has been one of the worst bottlenecks in the world. The situation steadily deteriorated following the Nigerian civil war; once the oil boom got underway the port found it increasingly difficult to handle the escalating volume of traffic. At the time of the July 1975 coup which brought Brigadier Mohammed to power, some 240 ships were waiting in the Lagos 'roads' to unload their cargoes and waiting time could be three or more months.

Originally Lagos was designed as a rail port, but Nigeria's antiquated railway system has long been unable to handle the volume of imports the country takes and 80 per cent is distributed by road. The problem results from Nigeria's oil boom and the consequent flood of imports; nor are her other harbours – Port Harcourt, Warri and Calabar – built on the scale to handle the volume. But, if Nigeria is to be the hub of a West African community and so the natural starting point for a regional communications system, her harbours must be modernized and their capacities enlarged. The Third Plan allocated N387 million to be spent on shipping, ports and inland waterways and the bulk of this will be for port improvement. There is an absolute limit to what Lagos can manage and Nigeria has to develop her other ports. Port Harcourt – the traditional gateway to the north as well as eastern Nigeria and the cheapest route out for Chad as well as handling Niger traffic – had still not in 1975 recovered its pre-civil war level of activity. Some of the problems to be tackled are technical, such as shallow drafts at the entrance to the Calabar river or the sandbar of the Escravos that is limiting developments in the Niger Delta at Warri and Burutu. These problems can be overcome; they need to be urgently. During 1975 as the crucial Third Plan got underway appalling delays were experienced in Nigeria simply because of congestion at the seaports and at the international airport at Ikeja. It was political necessity as well as justification for the coup that prompted the Mohammed regime to tackle the problem of the ports almost before anything else after replacing Gowon at the end of July 1975.

Lagos was laid out on an island to a colonial plan at a time when it was not envisaged either that it would have to handle the volume of traffic that was using it by the 1970s or that the city would burst its seams in terms of population. These facts coupled with the economic boom meant by 1975 that the city had become a byword in Africa for road chaos and traffic jams. These in turn immensely retarded the turn-round of vehicles and delivery in and out of the country's capital city so that – once more – the would-be starting point of the West African communications network is too small and inefficient to match actual

146

and potential needs.

Nigeria railways are a classic example of a colonial-built structure whose design is totally inadequate to the purposes of the post-independence era. Up to the early 1960s the railways – 2,178 miles of them – were the chief means of evacuating the produce of the northern states to the Nigerian coastal ports, but competition from the roads combined with neglect and ineffficiency made them daily less able to cope with the demands made upon them. The two main lines run from Lagos north to Zaria and beyond and from Port Harcourt similarly to the north: the system was never designed to link the country from east to west nor were branches extended to the eastern or western borders to link Nigeria with her neighbours. Apart from the geographic layout the lines were so poorly laid and subsequently maintained that derailment has become a modern feature of the system. The railway is to be changed from narrow to standard gauge as part of Nigeria's Third Plan.

The defects of Nigeria railways are obvious; the remedy a major operation. Two statistics will illustrate the railway's decline: agricultural freight handled in 1958–59 came to 850,000 tons; by 1970–71 it had declined to 350,000 tons. Passengers carried in 1961–62 totalled 11,000 people; in 1973–74 the figure was down to 4,670. Operational losses during the 1970s escalated: in 1972–73 they came to N21.8 million; the next year they were up to N33.1 million. The railway does not serve much of the heart of the country, especially many of the newly developing cities where commerce is rapidly expanding. For a country of Nigeria's vast size – 356,668 square miles – the railway is grossly inadequate; to make regional sense it needs to be extended north to Niger and Chad – a relatively simple matter – and, what is more important, it needs an entirely new east-west line to link it with its neighbours, Cameroon to the east and Dahomey to the west. Under her Third Plan Nigeria allocated N718.9 million for equipping the railway and constructing the new standard gauge. This should be regarded as no more than stage one of a modernization programme of which stage two ought to be a new east-west line.

Like most developing countries Nigeria devotes a great deal of attention in her plans to building up the country's infrastructure: not only is this especially needed as a result of the oil boom but, even more, an efficient communiations network centred upon Nigeria is a pre-requisite for the successful development of E C O W A S. It must be inevitable in economic terms that Nigeria becomes the focus of the West African Community and that in itself will carry enough political problems for any Nigerian government to handle. Since that is the case the onus rests particularly upon Nigeria to break the old colonial structural pattern of communications and start linking her system into those of her neighbours. Little planning (in this direction) has taken place in Nigeria since independence and even the programme set out in the ambitious Third Plan is likely to be swamped by the scale of the economic boom that oil has brought to the country. Communications, therefore, demand even greater priority of attention than they were accorded up to the mid-1970s.

An apparently ironic commentary upon the state of chaos of the Nigerian ports in 1975 was the announcement by two British firms that they were to take goods for Nigeria overland across the Sahara: the firms planned to use 12-metre trailers and expected the journey to last two or so weeks at a cost of £13,000 for each trailer. The story made headlines in the Nigerian press as a variant upon the constant tales of port congestion. Yet it highlighted a possible development of far greater value than simply to Nigeria. The Trans-Saharan route, properly used, could play a significant role in opening up southern Algeria and Niger as well as northern Nigeria while there could also be branches off to Dahomey (now Benin), Ghana and Upper Volta. Nigeria's need to overcome import problems in 1975 led her to adopt a number of expedients which could become permanent ways of developing a wider regional system of communications. Arrangements were also made in 1975 for Nigerian-bound ocean vessels carrying government consignments to discharge their cargoes at Ghanaian ports: subsequently these cargoes were to be sent overland to Nigeria. An agreement between the two governments was reached to do this and the Ghanaian Commissioner for Transport and Communications described the move as a concrete step towards the realization of the objectives of ECOWAS.

Nigeria's necessity has amply illustrated the inadequacies of the inherited colonial communications structures; in 1975 that necessity was being used as a starting point to bring about changed patterns of transport which, although initially designed to overcome purely Nigerian problems, could in the long run widen the scope of the regional structure. What the Nigerian oil boom made abundantly plain, as it produced the country's vast economic expansion, was the need for a revolution in its communications infrastructure: the trouble in this part of Africa is that few strategic communiations — whether roads, railways or rivers — either exist or, as yet, have been adequately developed.

Notes

1. See *West Africa,* June 1975.
2. See *Africa Magazine*, July 1975.
3. Nigerian *Sunday Times*, July 27th, 1975.

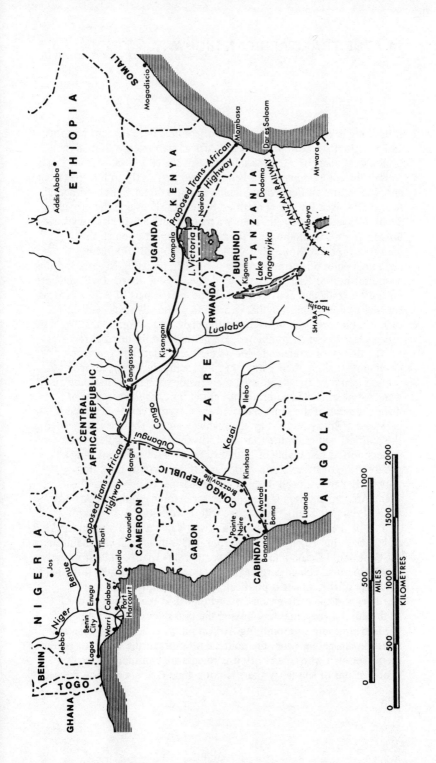

It has long been one of the leftovers of colonial development patterns and an irritant to both progress and African nationalism that anyone on the west coast of Africa wishing to visit the east coast — from Lagos to Nairobi, for example — has been obliged first to fly to Europe (Rome) and then return to East Africa. There were no direct trans-continental routes. This is at last being rectified by both airlines and the projected Trans-African Highway. In theory a major trans-continental highway linking Lagos and Mombasa will have great impact upon all the countries it passes through; it will link the two most flourishing economic poles of northern Africa — oil-rich, booming Nigeria, and the economic hub of East Africa, Kenya; in addition, the road will generate far greater trade and activity through the central part of Africa in areas that are now economically backward and politically neglected. In practice, however, the obstacles to the highway are formidable while its prospects, even when completed, are neither certain nor secure.

The theory of strategic highways changes little: they generate development along their paths; they link areas and so encourage both economic and political contacts; they ensure exchanges that previously were not possible and so bring into closer unity separate states or different areas of the same state. All these results can be achieved in part by a highway but not by the highway alone. There is also needed the political will to ensure that the highway does in practice what it is understood to be capable of in theory. The 1970s are witnessing a proliferation of highways in Africa and the growth of links that previously had not been deemed either feasible or desirable. The Trans-African Highway is regarded as a means of promoting African unity. How much unity in practice will it promote?

Part of the answer to this question will depend simply upon the kind of highway that emerges: a super four-lane tarmac highway slicing across the continent? Or an upgraded all-weather dirt road — at least in the middle section — since neither finance nor interest will take it further? And part of the answer is political. It was one thing for Robert Gardiner and the early enthusiasts who pushed the project in the E.C.A. to dream of a great highway linking the two sides of the continent for the first time and so promoting African unity. It becomes a different question altogether when the political interests of the different governments through whose territory it is to pass are considered. Despite protestations of solidarity there is not a great deal in common between the six countries — Nigeria, Cameroon, the Central African Republic, Zaire, Uganda and Kenya — through which the highway is to pass. Nigeria since the signing of the E C O W A S treaty in May 1975 has special reason to encourage the stretch of highway through Nigeria and Cameroon, although in economic terms she sees greater advantage to her interests in concentrating upon the Trans-Saharan Highway. At the other end the portion of the highway that passes through Kenya and Uganda already exists and conditions throughout the 1970s in the East African Common Market have been sufficiently difficult to discourage any high expectations that an extension of that highway into the centre of the continent will lead to any very spectacular increase in either trade or diplomatic harmony. The Central African Republic is so poor that, while it must benefit from the highway passing through its territory, its own contribution outwards in the form of increased trade and passage of its citizens is likely to be strictly limited. Zaire in the centre presents a rather special puzzle of its own. The northern areas of Zaire through which 1,000 miles of the highway must pass are in any case neglected and economically backward while communications between them and the capital, Kinshasa, are indifferent and slow. A trans-continental highway passing through this region and unconnected with Kinshasa could have the effect of detaching the region from Kinshasa still more than is currently the case, and fears of this may at least partly explain Mobutu's general reluctance to push the project. Such political considerations have to be matched against the economic possibilities.

The proposed route of the highway has long been in existence, although the actual roads are often of deplorable standards. The idea for a Trans-African Highway came from Japan in 1969 when a Japanese economic mission to Africa formulated a plan which was formally put to Kenya in 1970 by the Japanese Ambassador.[1] The original proposal was for a road passing through and linking seven countries — Kenya, Uganda, Zaire, the Central African Republic, Chad, Cameroon and Nigeria. It was to include 3,000 miles of macadamized four-lane highways. There was an active lobby in support of the proposal in Japan and the giant Mitsubishi Corporation was prepared to build the road. The Japanese government signified its willingness to give aid for the project. Such Japanese enthusiasm led to African suspicions: just what were Japanese intentions? From Tokyo's point of view there was much for Japan to gain from building the road: increased influence and trade on a continent where she sought ever greater quantities of raw materials for her rapidly expanding industries. Japan signified her willingness to work with any international organization.

In these early stages of the plan three of the countries — Kenya, Nigeria and Zaire — indicated their readiness to approve the Japanese idea. After further delays – in 1971 – the idea for the road was formally tabled by the Executive Committee of the E.C.A. in Addis Ababa. By then there were six countries involved, for Chad had dropped out. From these six – Nigeria, Cameroon, Central African Republic, Zaire, Uganda and Kenya – a Trans-African Highway Committee was drawn. There were many questions to answer: should the highway consist of a road, a railway, waterways or a combination of all these modes of transport? A roadway, if a new system of communications has to be established, is both the cheapest and quickest method of testing the economic needs of the area. As far as the proposed highway went the worst gaps to be faced were in Zaire. The E.C.A. Secretariat undertook responsibility for the first phase of the programme. It faced two questions: how viable was the proposal; and how far was it an ideological dream?

The Executive Secretary of the E.C.A., Robert Gardiner, established a Trans-African Highway Bureau with effect from July 1st, 1971. The route was to start from Mombasa in Kenya and then pass through Uganda, cut across the northern portions of Zaire, through the southern part of Central African Republic and Cameroon, through south-eastern Nigeria and finish at Lagos – a total of 4,400 miles. A road on such a scale, however, is slow to build, costly, and cannot easily be altered. As it was, the whole length of the proposed road already formed parts of the existing transport networks of the six countries so that it could be argued that those segments of the countries through which it was to pass were already considered essential communications areas by their individual governments. The road as it then existed consisted of 4,412 miles length, of which 3,497 miles were unimproved roads. In the first place additional work was to be restricted to 1,165 miles of the whole – that is, to the improvement of the poorest parts of the existing roads. The plan envisaged building new roads for only 5 per cent of the total distance. It was hoped to complete the highway during the Second Development Decade. A number of industrialized countries soon showed an interest in helping with the road. Three months after the Trans-African Highway Bureau had been set up, nine industrialized countries met on October 1st, 1971, and offered help. On October 25th, 1971, the British Overseas Development Administration put out a press release to say that the contract for the pre-feasibility study had been awarded to a London-based firm, T.P. O'Sullivan and Partners.[2]

The road itself faced a number of possibilities: there was a variant through northern Nigeria but that was dropped in favour of the route through the south and east where there is denser population and greater development. There were other variants through the north and south of the Central African Republic and through Zaire. It was far from easy to work out the road's potential. The original Japanese proposal was greatly scaled down though no precise figures were produced. Less than 10 per cent of the road actually has to be built; the rest exists, in some form or other. The early enthusiastic assessments — a macadamised road of four lanes — gave way to a more modest single-track all-weather road. Robert Gardiner said: 'We will build as cheaply, and also as efficiently as possible, and as the road begins to be more used (and to pay) we can upgrade. Thus we must determine minimum standards so that we can develop by stages.'[3] It was the voice of political and economic reason triumphing over earlier idealism and hope. By 1974, when small progress had been made, it was still hoped that the road would be completed by 1978 although Gardiner wanted it finished by 1976.

If the starting point was taken to be Mombasa then the first part of the road — Mombasa to Nairobi, through the Kenya Highlands to Eldoret and the Uganda border, across the Victoria Nile and to Kampala and then skirting the Ruwenzori moutains — was already virtually complete and in good high-grade condition. But the middle stretch faced far different conditions. The road has to pass through almost 1,000 miles of tropical jungle in northern Zaire through the Ituri forest and then from Kisangani northwards to cross the Zaire river. From Bangassou it crosses the Central African Republic to Bangui and on to Cameroon. In Cameroon the road is ill-defined, by-passing Douala and Yaounde for a route further north through Tibati and the Highlands. Finally it enters Nigeria and passes through Enugu, Onitsha, over the Niger, through Benin City and so to Lagos. The two ends of the road are in good condition: the Kenya-Uganda stretch is mainly asphalt and there is comparatively little upgrading to do; the Nigerian stretch is the same although parts in the east were damaged in the civil war and have not yet been fully repaired. The rest of the proposed road is in poor condition, however, and the 1,000 miles through Zaire are especially bad while parts in the Central African Republic and the Cameroon stretch in the Tibati region are not much better. In miles the road has to cover the following distances in the six countries: 570 miles in Kenya; 410 miles in Uganda; 980 miles in Zaire; 810 miles in the Central African Republic; 680 miles in Cameroon; and 550 miles in Nigeria. The three least developed parts of the road are in the three least developed and most inaccessible countries or parts of countries — Zaire, the Central African Republic and Cameroon.

The coming of the road must bring other reforms and co-ordination for the countries the highway passes through. Six main adjustments have to be implemented: all participants need to agree to allow commercial vehicles to collect and deliver freely across their frontiers; there have to be integrated traffic, vehicle and highway standards and regulations; foreign exchange facilities need to be established close to

153

isolated borders and road crossings; there ought to be as few immigration hurdles as possible and the abolition of visa requirements; efforts will be needed to ensure better frontier post efficiency and training for officials; and a 'green card' vehicle insurance system should be recognized between the countries.

By the end of 1971 concrete offers of aid for the road had been made. The Kenya section was to be surveyed by the British; Japan and Belgium were to survey the Zaire section; the section through the Central African Republic and Cameroon was to be covered by the West Germans and the French; and the Nigerian section was to be surveyed by the Americans. Between them a total of nine industrialized states — Italy, Netherlands and Sweden had joined the other six countries — had offered help while Greece was also demonstrating her desire to be involved. At the first meeting of the Trans-African Highway Committee in 1971, made up of senior public works officials and chaired by Zaire representative, Jean Mwenze, the chairman was able to report these various indications of international interest. Apart from the observers of the industrialized countries the meeting was also attended by officials from U.N.D.P., the African Development Bank, the World Food Programme, the World Bank, the International Road Federation and the Economic Commission for Asia and the Far East. As Robert Gardiner said: 'An encouraging aspect of this project is that it has caught the attention of businessmen and potential investors.'[4]

At the meeting of October 1st, 1971, the highway began to take shape at least in planning terms when the delegates of the nine industrialized countries indicated the extent of their support. Britain promised technical aid, the U.S.A. $50,000 for technical services, France technical assistance, while Japan, West Germany and Italy were willing to undertake further studies; Holland promised general technical assistance and $50,000; Belgium said her aid would be allocated to the Zaire stretch of the project. By early 1974 the costs were estimated at between £200 million and £300 million and there was need for a drastic scaling down of the project: grants and soft loans had been promised from Belgium, France, West Germany, Italy, Japan, the U.S.A., the African Development Bank and the World Bank; 50 per cent of the studies for the road had been completed, 40 per cent of these were assured of financial backing and 10 per cent were outstanding. The highway required more than the willingness of the countries concerned; it also needed the help and co-ordination of international organizations as well as aid from industrialized countries. The O.A.U. found it necessary to accuse the World Bank of a 'negative attitude' towards the highway and urged it to 'make its contribution'. The O.A.U. urged the six countries involved 'to mobilise all their human and material resources for the execution of the project'.[5]

154

In contrast to the T A N Z A M railway which Tanzania and Zambia tried in vain to interest the West in building, there has been no apparent lack of Western interest in the Trans-African Highway. Such interest in the new 'unity highway' must at least make Africans think, for whatever else the West wants in Africa it is uninterested in promoting unity. It can be argued that unlike the T A N Z A M this highway is not controversial from a Western point of view, while there are considerable prospects of new mineral discoveries in the huge central area it will pass through: when these are uncovered and exploited the existence of the road would mean swifter access to such discoveries, something no materials-hungry Western industrialized country is likely to quarrel about. Meanwhile, the problems faced by the six countries participating in the road are formidable. Little intra-African trade exists in the region and so far all economic studies have emphasized regional and subregional needs to the exclusion of continental developments. Once away from Kenya and Uganda in the east and Nigeria in the west, the whole area suffers from poor roads, empty shops and little active commerce. A key central stretch of 1,125 miles needed considerable improvement at once (in the words of the 1971 feasibility study) yet three years later nothing had been done. The most isolated regions of the Central African Republic and Zaire are economically backward with little immediate prospect of change; though exponents of the road argue that its coming must lead to economic advance in these remote areas, they seem to base this upon a hope rather than anything more precise. No studies have been carried out, for example, to show whether there would be increased demand in the Central African Republic or Northern Zaire for goods from either Nigeria or Kenya once the road was in effective operation.

The road also has to face political problems. Given the different nature of the six countries through which it passes, their varying regimes and the distinctive and individual styles of their rulers, there could easily enough be trouble for the international operation of the road should a major row erupt between any two adjoining countries through whose territory the road runs so that one or both insist upon closing their borders to each other. Rhodesia has already set a precedent for such behaviour; so has the E.A.C.M. with the temporary closing of the border roads between Tanzania and Kenya at the end of 1974. As Robert Gardiner has said: 'When you have a road crossing different countries, there may be periods of misunderstanding, and a decision not to allow vehicles to pass from one of the member states.'[6] This is surely an understatement. The consequence of a border closure after a pattern of traffic movement has been established could be dire. One result of such possibilities must be an examination of political and geographical alternatives: the road could go through Chad, Sudan,

Tanzania – but to mention the alternatives is itself tantamount to voicing suspicions. The suspicions, however, exist.

On the other hand, overcoming the problems and fears provides the justifications for the road. These can be sought under a number of headings: greater trade; greater movement of people and so greater unity and understanding; more economic growth not just between the countries concerned but also along the line of the road within each country itself; and an eventual sense of greater African strength, of an ability to act as one. Some of these justifications still require to be proven: the cultural impact, for example, or the political effect. The road will link three of the biggest and potentially strongest African markets: Nigeria, Zaire, and the East African Common Market. It will open up many agricultural areas and should generate employment both by its building – although that will be only transitory – and by the subsequent long-range provision of ancillary services that must go with such a highway.

A look in turn at the six countries shows clearly that the impact of the road on each of them will vary greatly. For Nigeria the road's impact could be least of all in relative terms since her own booming economy with its oil base will more than overshadow the effects of the highway; even so, the fact that Nigeria forms the starting point of one end of the highway is of great importance to the other countries along the route. The extent to which Nigeria is concerned, however, is problematic: her own economy is growing fast and she is feeling her rapidly developing power, yet almost all Nigeria's trade is external to Africa. And though Nigeria has accepted the project willingly enough, it is far from clear how much she will throw her political weight and enthusiasm behind it – and this is what the highway requires if it is to succeed.

Assessments of the possible and probable contributions to the highway of the other countries – and its effects upon them – are equally uncertain. What kind of impact the highway has upon Cameroon, for example, will not be obvious for some years: it is in parts of Cameroon that the road traverses areas where the average daily traffic density is no more than one vehicle.

In the case of the Central African Republic its distance from the sea and landlocked position as well as its lack of road and rail communications have ensured that it has a most uneven and retarded economy. Its one means of moving heavy freight is by the Ubangi river – a tributary of the Zaire: such freight has to travel 700 miles by barges between Bangui and Brazzaville whence it is then trans-shipped a further 300 miles by rail to Pointe Noire. It is difficult to see the Central African Republic's economy moving forward at any speed under such conditions; the road, therefore, could give the country a tremendous economic boost as well as helping bring it out of its general isolation. Much of the republic that the road will cross is savannah suitable for growing cotton and groundnuts and one third of the country's population lives in the area the road will traverse. A part of the road near Bangassou is exceptionally bad whilst too little traffic uses it to make government maintenance worthwhile. The republic could develop a wood industry oriented to Africa. In the case of this country the northern alternative for the road was found to be uneconomic, but whether and to what extent the current proposal will boost the economy remains to be seen. It certainly should to some extent. The Central African Republic's vulnerability as a result of its landlocked position has been driven home to it in the period since the oil crisis, for much of the country has simply been without fuel since then. The republic also has to face the fact that periodically the Ubangi dries up — it was too low for navigation for two months during 1974. By early 1975 the economy appeared to be verging on collapse. Shortages of all sorts were compounded by the lack of internal transport. The coming of the Trans-African Highway, therefore, could have substantial ameliorating effects upon such conditions.

Zaire has its own political and economic problems. The country is under the exceptionally personal rule of General Mobutu. The north-east of Zaire — the area the road will pass through — is the most agriculturally productive part of the country. It was badly disrupted during the crisis years of the first half of the 1960s. In 1969, however, it produced 44 per cent of the country's robusta coffee, 100 per cent of its arabica coffee, 100 per cent of its tea, 60 per cent of its rice, 35 per cent of its seed cotton, 30 per cent of its groundnuts, 25 per cent of its sugar, 20 per cent of its palm oil, 12 per cent of its rubber, 14 per cent of its fish, 80 per cent of its gold and 70 per cent of its tin. There is a Zaire road improvement programme — much of it for the northern region and much of that along the route of the Trans-African Highway. It had not got underway, however, at the start of 1975. There was no change a year later. The capital of the region, Kisangani, is a river port serving a vast forested area; yet here many of the roads are impassable during the long rainy season and fuel is in very short supply. This is the most productive and fertile region of Zaire and a good road could ensure food and labour movement for the whole country and would make industrial development elsewhere correspondingly easier as well as stimulating trade in the north and with the country's neighbours. At present, however, the roads in this region — between Kisangani and Kasindi — are some of the worst in Africa, and during the rains develop huge potholes and become quagmires. For the highway to succeed there must be a major political and economic initiative from Zaire: the north-east has been designated Zaire's 'third pole of industrial develop-

157

ment' yet the area remains neglected. There has been major Zaireaniz-
ation of the economy there and expatriate businessmen have been
replaced: too often, however, by favourites and party officials rather
than by people skilled in running economic and commercial affairs.

The Uganda case is different again. Uganda has far better developed
roads than the central territories on the route. Of three possible routes
the southern one for Uganda was chosen and the economic impact
should be considerable: the south is where the cotton growing of the
country is concentrated. At present, though, as a result of past history
and the line of the Uganda railway, the country suffers from a one-way
pull eastwards through Kenya to the coast. The road could play a
major part in offsetting this one-way economic trend and, instead,
could stimulate new westward development while at the same time
bringing some Zairean economic contacts and activity eastwards into
Uganda.

At the eastern end of the road Kenya is likely to benefit by some
major spin-offs. Overall, hers is the second most sophisticated economy
out of the six participating countries and she has always been the
dominant partner in the East African Common Market. A broadening of
possibilities for trade inwards as well as the chance for Mombasa to
become a larger port to handle at least some of Zaire's exports would
help increase Kenyan prosperity.

It is possible to predict many results from a project on the scale of the
Trans-African Highway, though until it is built they will be largely a
matter of surmise. In a sense the road embodies the concept of the
growth of poles — East Africa, northern Zaire and south-eastern Nigeria.
The East African Common Market especially should attract agricultural
goods from Zaire. Northern Zaire, while potentially very rich in
agriculture and minerals, is currently also very isolated; the greatest
stretch of the least improved road is in Zaire. Further, the increased
movement between the poles should have advantageous results for the
countries in between.

The road could well lead to a general increase in tourism and the
share in its benefits — for what they are worth. More Africans and non-
Africans could be encouraged to visit along the road and spend money:
the central countries especially should benefit from this although they
need tourist facilities. At present only Kenya, Uganda and perhaps the
Central African Republic are likely to benefit from the highway as a
route for international trade. In the other countries it serves mainly a
provincial function.

The Trans-African Highway has a long way to go before it starts
boosting international trade between its six members. Stretches of
the proposed road in Zaire and Cameroon in 1973 registered an average
of only one vehicle a day and in other stretches the ratio was between

one and ten. Only at the two ends was there a major difference: the Lagos end registered an average of 10,000 vehicles a day; the Mombasa end 6,000. The number fell rapidly thereafter as the inland journey proceeded. The growth of the road must depend upon how viable it is for each of the six countries it passes through. Any road improvement in Africa must be welcomed: it is bound to stimulate country to country trade and contact and so help change these economies from local subsistence to more sophisticated exchange economies while also facilitating the movement of perishable foodstuffs. It could play a vital role in fresh discoveries and subsequent exploitation of new mineral finds. And it will enable cash crops to be moved out more quickly and cheaply to the ports. The more the road is used the more will it come to be used and such use in itself will ensure that the road is extended and improved.

The Trans-African Highway can act as a growth network: it represents the co-ordinating of six national road projects into a web of all-weather links between the six countries. Eventually, too, it should also have extensions to the Trans-Saharan road from Algeria to Nigeria; to the Kenya-Ethiopian Highway; and possibly even further south through Tanzania and into Zambia. As Gardiner, the eternal optimist has said: 'We are opening up a whole continent, making the continent an entity instead of isolated portions – to foster a greater flow of trade and more intimate cultural and political relations among African states.'[7]

In April 1974, in Lagos, the Trans-African Highway Co-ordinating Committee had to admit that delays were occuring in the crucial 980 mile section through Zaire. No feasibility studies had been carried out for the Kisangani-Kasindi section although help for this had been promised by the Belgian government. The delays were in part due to recurring bickering between Kinshasa and Brussels and that year both sides were blaming each other and refusing to sign the draft terms of reference for the study.

The road's possibilities are important and potentially exciting. Yet despite this challenge there has been no noticeable enthusiasm either to build it or to make it work and this fact calls the whole project into question in a number of ways.

First, there does not appear to exist enough interest in any one of the countries concerned to give the road sufficiently high priority. This lack of concern stems from different causes from country to country. Nigeria, when she looks outwards from her own economic developments, is more interested in E C O W A S and the Trans-Saharan Highway. Kenya, at the other end, has had sufficient complications with the E.A.C.M. in recent years and is unlikely, therefore, to be an enthusiastic exponent of this new development. Although northern Zaire is of

159

immense potential that region's development problems appear remote to the concerns of the central government in Kinshasa. Although the road has to pass through some of the poorest and most isolated areas of both Cameroon and the Central African Republic where its coming should certainly make an impact, it probably has to come first and demonstrate its capacity to initiate change before it can engender enthusiasm as a project. In Uganda General Amin would seem to have other pre-occupations though he may take up the road at a future date.

Second, however appealing the idea in theory, in practice the road does not appear to be essential to any one of the countries concerned: they have other internal problems, other poles of development that appear likely to yield greater national gains.

Third, there is no pressing economic or political justification for the highway as there was in the case of the T A N Z A M railway between Zambia and Tanzania: none regards it as an essential lifeline.

Fourth, although all pay lip service to the ideal of African unity, each country in fact is sufficiently nationalistic and sufficiently wrapped up in its own development problems to regard the issue of the highway as a side issue, an agreeable international luxury rather than anything more important.

Already in the early planning days the highway was reduced from its proposed macadamized four lanes to a single all-weather road. That represented a considerable comedown from the early expectations no matter how bravely Mr Gardiner tried to make it sound only a pragmatic economic decision. It reflected more than economics; it reflected a good deal of political indifference as well.

The road, slowly, is coming into being. It has been pushed as an instrument of African unity by the idealists and, no doubt, can and will play a unifying role. There exists the danger that the idea has run ahead of economic and political realities on the spot in the various countries concerned, and the real danger exists that even when it has been completed — indifferently, perhaps, in parts — it will be inadequately maintained, it will disappoint in terms of promoting development and may as a result come to be regarded as an expensive white elephant.

The road, no doubt, will be completed; but for any political idea to work it has also to be coupled with a felt need and in the case of the Trans-African Highway it seems too much a case of the political idea preceding the economic and nationally-felt justifications for it.

Notes

1. Richard Taylor, 'Unity Highway', in *Africa Magazine*, No. 3, 1971.

160

2. *African Development*, January 1972, p. 55.
3. Richard Taylor, *op. cit.*
4. *ibid.*
5. Richard Synge, 'Trans-African Highway', in *Third World,* January/February 1975.
6. Richard Taylor, *op. cit.*
7. *ibid.*

PART THREE

15   CONCLUSION

The concept of strategic highways — roads, railways or rivers — whose unimpeded use is essential to a nation's well-being is as old as recorded history. The word 'strategic' has such powerful military connotations that equally important economic considerations can be overlooked. The African continent of the 1970s is witnessing an unprecedented growth of new highways alongside an even more urgent improvement of existing ones. The key motive in almost every case is economic development; yet always there is a political side to economic development and sometimes the political considerations are of overriding importance. There is a fashionable modern trend which argues that economics are separate from politics. They are not and they never have been. The constant twists in the story of communications in southern Africa in the decade following U.D.I. in Rhodesia should once and for all dispel the notion that politics and economics are not intertwined. And in that area at least the military connotations of the word strategic as applied to highways are only too obvious. To take on example: in March 1973 South Africa began to build an all-weather road from Grootfontein in Namibia into the Caprivi Strip whose basic purpose was to ensure easy access for her troops to an area that was coming increasingly under S W A P O guerrilla domination. By November 1975 she was using the road to take her troops to the Angolan border so that they could cross — at least into the southern parts of that country — in order to find and destroy S W A P O military camps.

It is important not to let the concept run away with realities. Distances in Africa are so vast that a theoretically strategic highway — such as the projected Trans-African Highway — may in practice turn out to be nothing of the sort: when completed this highway, whose justification is the promotion of greater African unity, could simply peter out in the middle of the continent through lack of interest and inadequate use though its proponents will hope for a quite different result. On the other hand, and more hopefully, the coming of any new means of transport in a continent where so much is still traditional can at once bring about change. When the first gleaming Chinese trains travelled the T A N Z A M railway in Tanzania empty carriages or trucks were simply taken over by crowds of local people with their goods for market, their livestock and their families as for the first time they found available a means of travel that would revolutionize existing patterns.

Politics and economics constantly battle for the ascendancy: this has certainly been the case as between Zambia and Rhodesia in the years 1965–75. For political reasons Smith closed the Rhodesian border with Zambia in January 1973. Within ten hours he realized the enormity of his economic mistake – he would deprive Rhodesia at that time of approximately £500,000 a month hard currency from railway freight charges for Zambian copper – so he announced that the closure excluded copper. Zambia ignored the exception and for two years managed to send her copper out along other routes. Then in November 1975 because politics and war in Angola had closed the Benguela railway to Zambian copper, Zambia was obliged again to route its copper through Rhodesia if only briefly.

There are fourteen landlocked countries in Africa and the state of their economies closely parallels the extent to which they have easy access to the sea; this in turn depends upon the political relationships they maintain with those neighbours through whose territory their key communications have to pass. The economic backwardness of a land-locked country like Chad can at once be appreciated when its incredibly poor communications outwards in any direction are understood. The relatively sophisticated and advanced, if lopsided, economy of Zambia can also be more easily understood when the number of her alternatives – however precarious they sometimes are – is recognized.

The attraction of poles is another concept closely aligned to strategic highways. The powerful economy of South Africa has constantly drawn southwards the economic links of countries like Rhodesia, Zambia, Malawi and Botswana despite major ideological differences and determined attempts to move away from her economic control. Maybe the 'pole' attractions of Nigeria and Kenya as the most sophisticated markets along the axis of the Trans-African Highway will help make the dream become a working reality.

Landlocked Zambia with its vast copper tonnages to export has developed into the nub of a network of strategic highways. Since her economy depends upon getting the copper out so have her politics centred upon the overriding question of communication: the TANZAM railway, another line through Angola, a fresh link to Malawi. Sometimes Zambia appears almost frantic in her determination to develop fresh communications alternatives and for her each one in turn takes on a strategic significance of its own. She may end up with more routes out than any other landlocked country anywhere but if she does so it will no more than reflect the traumatic early years of her independence when the peaceful developments she would have liked to pursue seemed to be threatened so often by the race tensions and politics of confrontation in that part of the continent.

Mozambique appears to be in a more than usually satisfactory position with regard to communications: she is on the sea and has three excellent deep-sea ports and is not dependent upon any other country for access to the sea or control of highways; instead, she holds the key to much of the strategic movement of the area since it is within her power to deny access to the sea to half a dozen inland countries. Yet here economics override purely political objectives, for Mozambique is so poor that she needs urgently the large revenues to be derived from South African and Rhodesian use of her railways and ports no matter how much she may abhor their policies. In consequence the Frelimo government of Samora Machel disappointed its outside ideological supporters in the months following the territory's independence in June 1975 because it did not at once − and could not afford to − close its borders to the white minority regimes though it was to close the border with Rhodesia early in 1976. The politics of communications were aptly illustrated by Mozambique when, on the eve of independence, Machel gave a bitter speech denouncing Banda of Malawi whom he accused of allowing the former Portuguese secret police to 'rob and swindle the Frelimo militants and kill the Frelimo military and the population'. He went on to remind Malawi of its absolute reliance upon Mozambique for access to the sea. Furthermore, even part of Malawi's inland trade could be disrupted by Mozambique should she decide to close the Blantyre to Salisbury road that crosses the Tete Province: then, for example, 13 per cent of Malawi's imports and 7 per cent of her exports would be affected.[1] At the same time that Mozambique was not immediately closing her borders to her white-ruled neighbours for ideological reasons, the mere possibility that she might do so was forcing basic reappraisals of policy in Pretoria and Salisbury which between them produced the détente exercise.

The military element in the communications story of southern Africa has always been extremely important. South African opposition to and fear of T A N Z A M was partly economic − that it would help detach Zambia from the South African economic sphere of control − and partly military: that in the event of a full-scale black-white military confrontation in the south the railway points like an arrow at the white heartland. Smith's closure of the border with Zambia in January 1973 was primarily for military reasons: to try to force Kaunda to control the Zimbabwe guerrillas who had then launched a highly effective campaign in the north-east of Rhodesia that threatened at the time to get beyond the control of the Rhodesian security forces. The closings of Lobito and the Benguela railway in 1975 were the direct result of military action in Angola and had dire economic and political consequences for Zambia. On his state visit to Britain in November 1975 President Nyerere of Tanzania made plain his scepticism of any fruitful outcome to the détente exercise.[2] He proved correct in his scepticism and black-white confrontation resumed in southern Africa in early 1976 so that once more the military significance of the various highways in the area became more important than some of their economic roles.

Zaire is so large and transport within her million square miles of territory so difficult that a new highway built entirely within that country can yet have a strategic significance whose effects reach far across its borders. The proposed railway linking Ilebo to Kinshasa falls into this category (see Chapter 11). When completed it will mean that for the first time effectively Shaba copper can be transported to the sea entirely through Zaire territory so that the country is no longer dependent upon the Benguela railway through Angola. More interesting and ironic, it will provide Zambia with yet another route to the sea and one that will be no longer than her new T A N Z A M route to Dar es Salaam.

Existing highways can be used to justify policies: this mainly was the Malawi rationalization for its pursuit of good relations with the white south contrary to the main stream of O.A.U. politics during the 1960s. Fear of an existing pattern warping a country's development can equally be used to justify the creation of a new highway as was the case with Zambia and the T A N Z A M railway. Communications form part of the lifeblood of a nation and highways can change in significance overnight. Military strategy, economic development, aspirations towards greater unity or the simple encouragement of more cultural exchanges are all mixed up with the pattern of a country's communications with its neighbours: no area of the world demonstrates all these aspects of communications so clearly or with so many exciting if dangerous possibilities as Africa in the middle of the 1970s.

Notes

1. See *African Development*, October 1975, p. 85.
2. See *The Times*, November 24th, 1975 (leader).

African Development A valuable source of economic data
Africa Magazine Numerous issues over the period 1971–75
Birmingham, W., and A.G. Ford, *Planning and Growth in Rich and Poor Countries* (George Allen & Unwin, London, 1965)
Bixler, R.W., *Anglo-German Imperialism in South Africa 1880–1900* (Warwick & York, London, 1932)
Bruwer, J.P. van S., *South West Africa: The Disputed Land* (Nasionale Boekhandel Bpk, Cape Town, 1966)
Cervenka, Zdenek (ed.), *Landlocked Countries of Africa* (The Scandinavian Institute of African Studies, 1973)
Elliott, Charles (ed.), *Development of Zambia* (O.U.P., London, 1971)
First, Ruth, *South West Africa* (Penguin, Harmondsworth, 1963)
Fordham, Paul, *The Geography of African Affairs* (Penguin, Harmondsworth, 1974: 4th ed.)
Green, R.H., and A. Seidman, *Unity or Poverty?* (Penguin, Harmondsworth, 1968)
Griffiths, I.L., *Transport and Communications in the Relationship of a Land-Locked State: Zambia* (paper), Transport in Africa (Centre of African Studies, University of Edinburgh, 1969)
Hailey, Lord, *An African Survey* (O.U.P., London, 1956: revised)
Hall, R., *Stanley* (Collins, London, 1974)
Hance, W.A., *African Economic Development* (Pall Mall Press, London, 1967)
Hazlewood, A. (ed.), *African Integration and Disintegration* (O.U.P., Oxford, 1967)
Hutchinson, R., and G. Martelli, *Robert's People* (The Life of Sir Robert Williams Bart 1860–1938) (Chatto & Windus, London, 1971)
Jeune Afrique, *Atlas Afrique* (Jeune Afrique, 1972)
Katzenellenbogen, S.E., *Railways and the Copper Mines of Katanga*, (Clarendon Press, Oxford, 1973)
Langer, W.L., *The Diplomacy of Imperialism* (Alfred A. Knopf, New York, 1935: reprinted 1968)
Legum, C. (ed.), *Africa Contemporary Record*, for the years 1970–75, (Rex Collings, London)
Lewin, E., *The Germans and Africa* (Cassell, London, 1915)
Listowel, J., *The Other Livingstone* (Julian Friedmann, London, 1974)
McMaster, C., *Malawi – Foreign Policy and Development*, (Julian

Friedmann, London, 1974)

O'Connor, A.M., *Recent Railway Construction in Tropical Africa* (paper), Transport in Africa (Centre of African Studies, University of Edinburgh, 1969)

Parsons, Q.N., *Economics of the Zambia-Botswana Highway, Enterprise*, No. 3, 1974 (Publicity & Information Department, Z.I.M.C.O.)

Perham, Margery, *Lugard, The Years of Adventure 1858–98*, (Collins, London, 1956)

Pettman, J., *Zambia: Security and Conflict* (Julian Friedmann, London, 1974)

Robinson, R. and J. Gallagher with A. Denny, *Africa and the Victorians* (Macmillan, London, 1961)

Short, P., *Banda* (Routledge & Kegan Paul, London, 1974)

Sillery, A., *Botswana: A Short Political History* (Methuen, London, 1974)

South Africa 1974, an official handbook of the Republic of South Africa

Synge, R., *Trans-Africa Highway, Third World*, January/February issue, 1975

White, H.P. and M.B. Gleave, *An Economic Geography of West Africa* (G. Bell, London, 1971)

INDEX

Abidjan, 144
Accra, 139
Addis Ababa, 11, 152
African Development Bank, 124, 154
African Maritime Conference, 144
African National Congress (A.N.C.),
    22, 64, 101, 110
Agip, 78
Alaska Air, 117
Algeria, 145, 148, 159
Algiers, 19
Alvor, 104
Amatongaland, 27
Amax see American Metal Climax
Amazon, River, 132
American Metal Climax (Amax), 84,
    100
Amin, Idi, 9, 10, 18, 160
A.N.C. see African National Congress
Anglo-American Corporation, 84,
    100-1
Angola, guerrilla war in, 9, 16, 20-1,
    39, 50, 81, 101-6, 110-11, 136-
    7, 163-4; relations with Zambia,
    13, 63, 67-70, 73, 76, 80-2, 85,
    93, 101-6, 116; emigration of
    labour to South Africa from, 39;
    strategic importance of Zambezi
    to, 47, 52, 55-6; road systems in,
    93; importance of Benguela railway
    to, 93, 97-100, 106, 126; develop-
    ment of railway system in, 93-
    102, 163; relations with Zaire,
    103-6, 129, 131-2, 135-7, 165
Angola Agreement (1975), 104
Antwerp, 93
Arnot power station, 31
Atlantic Ocean, 28-9, 68, 85, 93-4,
    100, 129-30, 133, 135
Australia, 31

Bahamas, 95
Balaka, 83
Banana, 132
Banda, Dr Hastings, 68, 82-3, 87-
    92, 99, 164
Bangassou, 153, 157
Bangui, 133-4, 153, 156
Barotse plain, 48, 53
Barotseland, 39, 47, 55; see also
    Zambia
BasCongo-Katanga line, 131
Bastos, Dr Pereira, 88
Basutoland, 34; see also Lesotho
Batoka plateau, 39
Bechuanaland, 25, 34-7, 55, 60, 72,
    82, 94; see also Botswana

Bechuanaland Extension Bill, 36
Bechuanaland Railway Company, 26
Beira, Botswana access to, 38; British
    naval blockade of, 112; com-
    petition for Benguela railway, 93,
    115; facilities at, 109; Malawi's
    access to, 87-9; rail link with
    Chindio, 54; Rhodesian access to,
    17, 19-20, 29, 60-3, 74, 108-9,
    112-13; road links with, 108-9;
    Zaire's access to, 72, 109, 131,
    135; Zambian access to, 13, 21,
    62-3, 67, 69-70, 73, 76, 78,
    82-3, 104, 110, 116-17; see also
    Blantyre-Beira line
Beira Convention (1950), 87
Beit, Alfred, 37
Beit Bridge, 74, 116; see also Rutenga-
    Beit Bridge line
Belgium, 154, 159; see also Congo,
    Belgium
Benguela railway; political and
    strategic significance of, 12, 19,
    60, 93-7, 106; importance to
    Zambia of, 13, 16, 21, 68, 70, 73,
    75-6, 78, 81-5, 93-4, 98-106,
    126; link with Cape Railway, 28;
    Cubal Variant on, 78, 81, 93, 98,
    121; route of, 93-9; importance
    to Zaire of, 93-100, 103-5, 129-
    31, 135-7, 165; traffic on, 93, 98;
    future rail links with, 93, 99-102,
    104; development of, 93-100;
    importance to Angola of, 93, 97-
    100, 106, 126; effects of civil
    strife in Angola on, 101-6, 110,
    163-4; competition with other
    railway lines, 115
Benguela Railway Company, 75, 93;
    see also C.F.B.
Benin, 148; see also Dahomey
Benin City, 153
Benue, River 140
Biafra, 17
Blantyre, 13, 88-9, 164
Blantyre-Beira line, 83, 87, 108
Bobdo, 131
Boers, 25, 27, 30, 35-7, 71
Boma, 131; see also Tshela-Boma line
Botswana, Rhodesian rail outlet to sea
    through, 9, 20, 37-8, 60-2, 112;
    and UN sanctions against
    Rhodesia, 9; relations with South
    Africa, 9-10, 13, 30, 32, 34-46,
    114, 163; planning of Trans-East
    African highway, 11; development
    of Cape Railway through, 12, 28,

168

Grey, George, 95
Grey, Sir Edward, 95
Grootfontein, 9, 56, 162
Grove International, 41
Groveput, 31
Gwelo, 60–1, 89, 108
Gwembe Valley, 77

Hailey, Lord, 39
Hazlehurst, Peter, 116
Heligoland, 55–6
'Hell Run', 73–9, 116–7, 121
Hofmeyr, J. H., 25–6
Holub, 39

I.B.R.D. (International Bank for
    Reconstruction and Development),
    123
Ifakara, 124
Ikeja, 146
Ilebo, 60, 131–3, 135, 165; see also
    Port Francqui
Imperial British East Africa Company,
    14
Indeco, 77–8, 126
Indian Ocean, 29, 47, 50, 55, 88, 100,
    108, 114, 135
Industrial Development Corporation
    Limited, see Indeco
Inga dam, 132
International Court at the Hague, 42
International Road Federation, 154
Israel, 10
Italy, 154
Ituri forest, 153

Jackson, Sir R., 78–9
Japan, 132, 151, 153–4
Jebba, 17
Johannesburg, 10, 24–8, 36–7, 108,
    117
Johannesburg Star, 58
Johnston, Harry, 26
Jonathan, Lebua, 22
Jos, 134

Kabalo, 131
Kabwe, 71–7, 88, 95; see also Broken
    Hill
Kafue, 77
Kafue Gorge hydro-electric scheme,
    48, 57, 79
Kainji dam, 141
Kalahari Desert, 27
Kalomo, 77
Kamina, 97
Kampala, 10, 73, 153
Kanem, 143
Kano, 134
Kansanshi mine, 84, 95–6, 100–1

Kansanshi Copper Mine Limited, 100
Kapiri Mposhi, 10, 12, 13, 73, 77,
    116, 119–20
Kariba dam, 47–9, 57–8, 69, 71,
    73–4, 79
Kariba, Lake, 47, 53, 57
Kasai, 133
Kasai River, 97, 131–2
Kasindi, 157, 159
Katanga, 60, 71–3, 93–7, 100, 115,
    129–131, 137; see also Shraba;
    BasCongo-Katanga line
Katanga-Dilolo-Lubumbashi line;
    see K.D.L.
Katima Mulilo, 39, 41, 43–4, 52, 56
Kaunda, Kenneth, development of
    Zambian railway system, 13, 83–
    5, 88, 93, 99, 106, 115, 118–21,
    123–5; meeting with Smith and
    Vorster (August 1975), 15;
    detente with South Africa, 16, 21;
    building of BOTZAM road, 44;
    closure of border with Rhodesia
    (1973), 62–3, 78, 164; relations
    with Banda, 83, 88, 90, 99;
    relations with Nyerere, 88, 102,
    121, 136; and Zambian economic
    development, 100; relations with
    Mobutu, 136
Kavango, 56; see also Caprivi Strip
Kazungula ferry, 13, 39–43, 45–8,
    52, 57, 67
K.D.L. (Katanga-Dilolo-Lubumbashi
    line), 81, 94, 99, 102
Kelemie, 131
Kenya, relations with Uganda, 9, 10,
    18; transport links with Zambia,
    68, 71, 126; activities of
    Tanganyika Concessions (Tanks)
    in, 95; British administration of,
    116; closure of border with
    Tanzania (December 1974), 126,
    155; importance of Trans-
    African highway to, 150–5, 158–9,
    163
Kenyatta, Jomo, 18
Khama, Seretse, 9, 38, 44, 60, 114,
    136
Kigoma, 131, 135
Kilombero valley, 125
Kimberley, 24, 26, 27, 95
Kindu, 130–1
Kinshasa, 103, 130–3, 135, 151,
    160, 165
Kinshasa-Matadi line, 131, 133
Kisangani, 130–1, 133, 153, 157,
    159
Kissinger, Henry, 40
Kitwe, 77
Knutsford, 1st Viscount, 26, 36

Rhodesia (March 1976), 9, 17, 20–2, 29, 51, 62–3, 111, 113–4, 164; rail links with Tanzania, 13; achievement of independence by, 15, 49, 52, 61, 110–4; relations with South Africa, 21, 28–30, 32, 49–50, 52–3, 58–9, 62, 80, 108–14, 164; development of railway system in, 24–7, 60, 64, 108–9, 135; relations with Swaziland, 34–5, 108, 112, 114; guerrilla war in, 39, 48–50, 52, 57–8, 61, 63, 90–1, 112; strategic importance of Zambezi to, 47–50, 52–3, 57–8, 108; Cabora Bassa dam in, 49–50, 58–9; relations with Zambia, 67–70, 74, 80, 104, 108–14, 116; communications with Malawi, 87–92, 108–9, 114, 164; construction of Nacala line, 87–92, 108–9; involvement in Angolan civil war, 105; port facilities in, 108–14, 127, 164; road system in, 108–9; strategic and political power of, 108–14, 164; airports in, 108
Mozambique Convention (1909), 108
Mpika, 73, 77, 125
Mpimbe, 89
M.P.L.A. (Popular Movement for the Liberation of Angola), 69, 99, 101, 103–5, 137
Msonthi, J., 88
Mtwara, 67, 87–8
Mufulira, 77
Muller, Dr. H., 44
Munbere, 131
Mwenze, Jean, 154
Mwenzo, 126
Mwinilunga Exploration (1970) Company, 84

Nacala, 13, 67, 109, 135
Nacala line, 68, 76, 78, 82–3, 87–92, 104, 108–9
Nairobi, 10, 150, 153
Namibia, and Grootfontein-Caprivi Strip road, 9, 162; South Africa's relationship with, 20–1, 30, 44, 52–3, 111; railway lines in, 28; boundary with Cape Province, 29; links with Botswana, 34, 38, 42, 45–6; future independence of, 35, 38, 42, 45–6; strategic significance of Zambezi to, 49, 57; guerrilla war in, 57; geographical proximity to Zambia, 67; see also South West Africa
Nampula, 108
Nassau, 95

Nata, 39–43, 45
Natal, 24–8, 31
National Front for the Liberation of Angola, see F.N.L.A.
National Transport Corporation (Zambia), 81
Nchanga Consolidated Copper Mines (N.C.C.M.), 84, 101, 117
Ndola, 46, 76–7, 82, 94, 97, 100, 121
Netherlands, 154
Netherlands Railway Company, 24
Neto, Dr A. A., 101, 103, 106, 132
Ngaliema, 132; see also Stanley Falls
Ngamiland, 39
Ngoma, 41, 43–4
Ngonye Falls, 53
Ngwenya, 35, 112
Niger, 13, 140–1, 146–8
Niger, River, 17, 140–1, 146, 153
Niger River and Sea Transport Company (S.N.T.F.M.), 140
Nigeria, railway development in, 14–15, 18, 145–7; dominant role in E.C.O.W.A.S., 15, 139, 142–3, 147–8, 151, 159; civil war in, 17, 146, 153; and Trans-African highway, 19, 139, 145, 150–6, 158–9, 163; and Trans-Saharan highway, 19, 139, 145, 148, 151, 159; communications with Chad, 134, 146–7; congested ports of, 139, 145–6, 148; waterways in, 139–41, 145–6; road system in, 145, 147, 153; airways in, 145–6
Nile, River, 14
Nixon, Richard, 40
Nkomo, Joshua, 110
Northern Rhodesia, 48, 58, 60, 64, 70–4, 82, 88, 94–7, 115–6, 118, 130; see also Zambia
Nova Freixo, 89
Nova Lisboa, 98–9
Nyasa, 26, 54
Nyasa, Lake, 87–8
Nyasaland, 54, 58, 70, 87–9, 109, 116; see also Malawi
Nyerere, Julius, relations with Amin, 18; relations with Banda, 87; relations with Kaunda, 88, 102, 121, 136; attitude to TANZAM railway, 120–4; attitude to China, 125; relations with Mobutu, 136; and détente in Southern Africa, 164–5

O.A.U., policy towards white minority regimes, 17, 30, 68, 82, 91, 136, 165; Banda's relations with, 91; relations with E.C.O.W.A.S., 144;

173

175

Vungu Vungu Project, 56

Walvis Bay, 29, 34, 38, 42
Wankie, 69, 72, 95, 97
Wankie Game Reserve, 40
Warri, 146
Welensky, Roy, 57
West Germany, 97, 123, 154
West Nicholson, 62
Williams, Sir Robert, 95–7, 99
Wina, Arthur, 110
Witwatersrand, 22, 24–7, 32, 36, 39, 108
Witwatersrand Labour Association, 39
Wonderfontein, 31
World Bank, 74, 83, 123–4, 154
World Food Programme, 154

Yaounde, 134, 153
Yaounde agreements, 143

Zaire, and Kenyan/Ugandan crisis (1976), 11; relations with Zambia, 13, 67, 70, 72–3, 75, 83, 85, 101–2, 129–32, 135–6; railway system in, 15, 82, 94–100, 102, 130–3, 135–7, 165; trade with South Africa, 30, 32, 131, 135–6; trade with Botswana, 45; rail links with Rhodesia, 60–1, 72; important mineral deposits in, 69, 93, 129–32, 135; access to Beira, 72, 109, 131, 135; importance of Benguela railway to, 93–100, 103–6, 129–31, 135–7, 165; relations with guerrillas in Angola, 103–6, 135–7; relations with guerrillas in Mozambique, 104; river system in, 129–35; communications with Tanzania, 129, 131, 135; and Trans-African Highway, 129, 135–6, 151–61; road system in, 130, 135–6, 157; relations with China, 132; relations with Rhodesia, 135–6
Zaire, River, 15, 129–35, 153, 156; see also Congo, River
Zambesia (Zambezia) 26, 28, 54; see also Zambia
Zambesia Consolidated Finance (Z.C.F.), 96
Zambesia Exploration Company, 84, 95–6
Zambezi (River), Kazangula ferry over, 13, 39, 43–4, 47,–8, 57; failure as highway, 15, 47–8, 53–5, 59, 67, 108; as strategic divide, 15, 48–50, 52–3, 55–9, 71; South African troop dispositions

on, 22, 39, 52–3, 56–8; as frontier between 'black' and 'white' Africa, 42–3, 47–50, 52–3, 57–9, 62–3, 79, 119, 136; Cape colonists expansion to, 25; Portuguese advance up, 26; development of Cape railway to, 28; geography of, 47; as source of power, 47–50, 53, 57–9; history of Caprivi Strip, 54–8; as frontier between Malawi and Mozambique, 88
Zambia, construction of TANZAM railway, 9–10, 119, 124–7; planning of Trans-East African highway, 11, 155; political significance of TANZAM railway, 12–13, 32, 73, 77, 81, 85, 115–27, 160; relations with Zaire, 13, 67, 70, 72–3, 75, 83, 85, 101–2, 129–32, 135–6; political significance of BOTZAM road, 13, 35, 39–46; transport links with Malawi, 13, 67–8, 73, 76, 78, 82–3, 88, 90–2, 99, 104, 125; relations with Angola, 13, 63, 67–70, 73, 76, 80–2, 85, 93, 101–6, 116; relations with South Africa, 13, 16, 21, 30, 32, 35, 56, 68–71, 73, 78–80, 117, 120, 124, 127; relations with Rhodesia, 14–15, 21, 39, 41, 45, 47–9, 57–8, 60–5, 67–82, 97, 106, 115–27, 163–4; strategic importance of Zambezi, 15, 47–9, 52, 56–8, 62; copper exports, 16, 21, 28, 63, 67–70, 75–6, 78, 80–5; importance of Benguela railway to, 16, 21, 28, 68, 70, 73, 75–6, 81–5, 93–4, 98–106, 126, 164; river links with Botswana, 34, 38–44, 47–8, 57–8, 67; development of BOTZAM road, 39–46; economic significance of BOTZAM road, 42, 45–6; development of Kafue gorge, 48, 57, 79; military resources of, 52; development of Kariba dam, 57, 69, 73–4, 79; access to ports, 67–70, 72–85; relations with Mozambique, 67–70, 74, 80, 104, 108–14, 116; need to diversify transport links, 67–73, 76, 85, 87, 106, 112, 115–16, 120, 126, 163, 165; capacity of TANZAM railway, 68; relations with Portugal, 68–71, 73, 80–1, 120; transport links with Kenya, 68; development of railway system in, 71–7, 81–5, 93–4, 98–102, 115–19, 124–7; road system in,

177